SpringerBriefs in Computer Science

For further volumes:
http://www.springer.com/series/10028

Min Chen • Shiwen Mao • Yin Zhang
Victor C.M. Leung

Big Data

Related Technologies, Challenges and Future Prospects

Min Chen
School of Computer Science
and Technology
Huazhong University of Science
and Technology
Wuhan, China

Yin Zhang
School of Computer Science
and Technology
Huazhong University of Science
and Technology
Wuhan, China

Shiwen Mao
Auburn University
Auburn, AL, USA

Victor C.M. Leung
Electrical and Computer Engineering
The University of British Columbia
Vancouver, BC
Canada

ISSN 2191-5768 ISSN 2191-5776 (electronic)
ISBN 978-3-319-06244-0 ISBN 978-3-319-06245-7 (eBook)
DOI 10.1007/978-3-319-06245-7
Springer Cham Heidelberg New York Dordrecht London

Library of Congress Control Number: 2014937319

Printed on acid-free paper

Springer is part of Springer Science+Business Media (www.springer.com)

Preface

"How big is big?" Science writer Stephen Strauss asks in his fun book for kids titled *How Big is Big* and explains that "bigness is something no one can consume."

In this book, we aim to answer this interesting question, but in the context of computer data. In the *big data* era, we are dealing with the explosive increase of global data and enormous datasets. Unlike seemingly similar terms such as "massive data" or "very big data," *big data* refers to the datasets that could not be perceived, acquired, managed, and processed by traditional Information Technology (IT) and software/hardware tools within a tolerable time. It can be characterized by four Vs, i.e., Volume (great volume), Variety (various modalities), Velocity (rapid generation), and Value (huge value but very low density).

In this book, we provide a comprehensive overview of the background and related technologies, challenges and future prospects of big data. We first introduce the general background of big data and review related technologies, such as cloud computing, Internet of Things (IoT), data centers, and Hadoop. We then focus on the four phases of the value chain of big data, i.e., data generation, data acquisition, data storage, and data analysis. For each phase, we introduce the general background, discuss the technical challenges, and review the latest advances. We next examine the several representative applications of big data, including enterprise management, IoT, online social networks, healthcare and medical applications, collective intelligence, and smart grid. This book is concluded with a discussion of open problems and future directions. We aim to provide the readers a comprehensive overview and big-picture of this exciting area. We hope this monograph could be a useful reference for graduate students and professionals in related fields, and general readers who will benefit from an understanding of the big data field.

We are grateful to Dr. Xuemin (Sherman) Shen, the SpringerBriefs Series Editor on Wireless Communications. This book would not be possible without his kind support during the process. Thanks also to the Springer Editors and Staff, all of whom did their usual excellent job in getting this monograph published.

This work was supported by China National Natural Science Foundation (No. 61300224), the Ministry of Science and Technology (MOST), China, the International Science and Technology Collaboration Program (Project No.:

2014DFT10070), and the Hubei Provincial Key Project (No. 2013CFA051). Shiwen Mao's research is supported in part by the US National Science Foundation (NSF) under Grants CNS-1320664, CNS-1247955, CNS-0953513, and DUE-1044021, and through the NSF Broadband Wireless Access & Applications Center (BWAC) Site at Auburn University (NSF Grant IIP-1266036). The research of Victor Leung is supported by the Canadian Natural Sciences and Engineering Research Council, BC Innovation Council, Qatar Research Foundation, TELUS, and other industrial partners. Any opinions, findings, and conclusions or recommendations expressed in this material are those of the authors and do not necessarily reflect the views of the foundation.

Wuhan, China — Min Chen
Auburn, AL — Shiwen Mao
Wuhan, China — Yin Zhang
Vancouver, BC, Canada — Victor C.M. Leung
January 2014

Contents

Acronyms

AMI	Advanced Metering Infrastructure
APT	Advanced Persistent Threat
BI	Business Intelligence
BLOB	Binary Large Object or Basic Large Object
BPM	Business Process Management
BSON	Binary JSON
CEO	Chief Executive Officers
CIO	Chief Information Officer
DAS	Direct Attached Storage
DMA	Direct Memory Access
ETL	Extract, Transform and Load
ERCIM	European Research Consortium for Informatics and Mathematics
GUI	Graphic User Interface
HDFS	Hadoop Distributed File System
HGP	Human Genome Project
HQL	HyperTable Query Language
ICT	Information and Communications Technology
IDC	International Data Corporation
IoT	Internet of Things
IT	Information Technology
LHC	Large Hadron Collider
Libpcap	Packet Capture Library
MMF	Multi-Mode Fiber
MPI	Message Passing Interface
MR	MapReduce
MVCC	Muti-Version Concurrency Control
NAS	Network Attached Storage
NER	Named Entity Recognition
NIST	National Institute of Standards and Technology
NLP	Natural Language Processing
NSF	National Science Foundation

OFDM	Orthogonal Frequency-Division Multiplexing
OLAP	On-Line Analytical Processing
OpenMP	Open Multi-Processing
PB	Petabyte
PMU	Phasor Measurement Unit
PNUTS	Platform for Nimble Universal Table Storage
RAID	Redundant Array of Independent Disks
RDBMS	Relational Database Management System
SAN	Storage Area Network
SDK	Software Development Kit
SDSS	Sloan Digital Sky Survey
SNS	Social Networking Services
SSD	Solid-State Drive
TB	Terabyte
TOMS	Topic-oriented Multimedia Summarization System
TOR	Top Rack Switches
URL	Uniform Resource Locator
WDM	Wavelength Division Multiplexing
ZC	Zero-copy

Chapter 1
Introduction

Abstract The term of *big data* was coined under the explosive increase of global data and was mainly used to describe these enormous datasets. In this chapter, we introduce the definition of big data, and review its evolution in the past 20 years. In particular, we introduce the defining features of big data, as well as its 4Vs characteristics, including Volume, Variety, Velocity, and Value. The challenges brought about by big data is also examined in this chapter.

1.1 Dawn of the Big Data Era

Over the past 20 years, data has increased in a large scale in various fields. According to a report from International Data Corporation (IDC), in 2011, the overall created and copied data volume in the world was 1.8ZB ($\approx 10^{21}$B), which has increased by nearly nine times within 5 years [1]. Such figure will double at least every other 2 years in the near future.

The term of *big data* was coined under the explosive increase of global data and was mainly used to describe these enormous datasets. Compared with traditional datasets, big data generally includes masses of unstructured data that need more real-time analysis. In addition, big data also brings new opportunities for discovering new values, helps us to gain an in-depth understanding of the hidden values, and incurs new challenges, e.g., on how to effectively organize and manage such data. At present, big data has attracted considerable interest from industry, academia, and government agencies. For example, issues on big data are often covered in public media, including *The Economist* [2, 3], *New York Times* [4], and *National Public Radio* [5, 6]. Two premier scientific journals, *Nature* and *Science*, also started special columns to discuss the importance and challenges of big data [7, 8]. Many government agencies announced major plans to accelerate big data research and applications [9], and industries also become interested in the high potential of big data recently. The era of big data is coming beyond all doubt [12].

M. Chen et al., *Big Data: Related Technologies, Challenges and Future Prospects*,
SpringerBriefs in Computer Science, DOI 10.1007/978-3-319-06245-7_1,

Recently, the rapid growth of big data mainly comes from people's daily life, especially related to the service of Internet companies. For example, Google processes data of hundreds of PB and Facebook generates log data of over 10 Petabyte (PB) per month; Baidu, a Chinese company, processes data of tens of PB and Taobao, a subsidiary of Alibaba, generates data of tens of Terabyte (TB) on online trading per day. While the amount of large datasets is drastically rising, it also brings about many challenging problems demanding prompt solutions. First, the latest advances of information technology (IT) make it more easily to generate data. For example, on average, 72 h of videos are uploaded to YouTube in every minute [13]. Therefore, we are confronted with the main challenge of collecting and integrating massive data from widely distributed data sources. Second, the collected data is increasingly growing, which causes a problem of how to store and manage such huge, heterogeneous datasets with moderate requirements on hardware and software infrastructure. Third, in consideration of the heterogeneity, scalability, real-time, complexity, and privacy of big data, we shall effectively "mine" the datasets at different levels with analysis, modeling, visualization, forecast, and optimization techniques, so as to reveal its intrinsic property and improve decision making.

The rapid growth of cloud computing and the Internet of Things (IoT) further promote the sharp growth of data. Cloud computing provides safeguarding, access sites, and channels for data asset. In the paradigm of IoT, sensors all over the world are collecting and transmitting data which will be stored and processed in the cloud. Such data in both quantity and mutual relations will far surpass the capacities of the IT architectures and infrastructure of existing enterprises, and its realtime requirement will greatly stress the available computing capacity. Figure 1.1 illustrates the boom of the global data volume.

1.2 Definition and Features of Big Data

Big data is an abstract concept. Apart from masses of data, it also has some other features, which determine the difference between itself and "massive data" or "very big data." At present, although the importance of big data has been generally recognized, people still have different opinions on its definition. In general, big data refers to the datasets that could not be perceived, acquired, managed, and processed by traditional IT and software/hardware tools within a tolerable time. Because of different concerns, scientific and technological enterprises, research scholars, data analysts, and technical practitioners have different definitions of big data. The following definitions may help us have a better understanding on the profound social, economic, and technological connotations of big data.

In 2010, Apache Hadoop defined big data as "datasets which could not be captured, managed, and processed by general computers within an acceptable scope." On the basis of this definition, in May 2011, McKinsey & Company, a global consulting agency announced Big Data as "the Next Frontier for Innovation, Competition, and Productivity." Big data shall mean such datasets which could

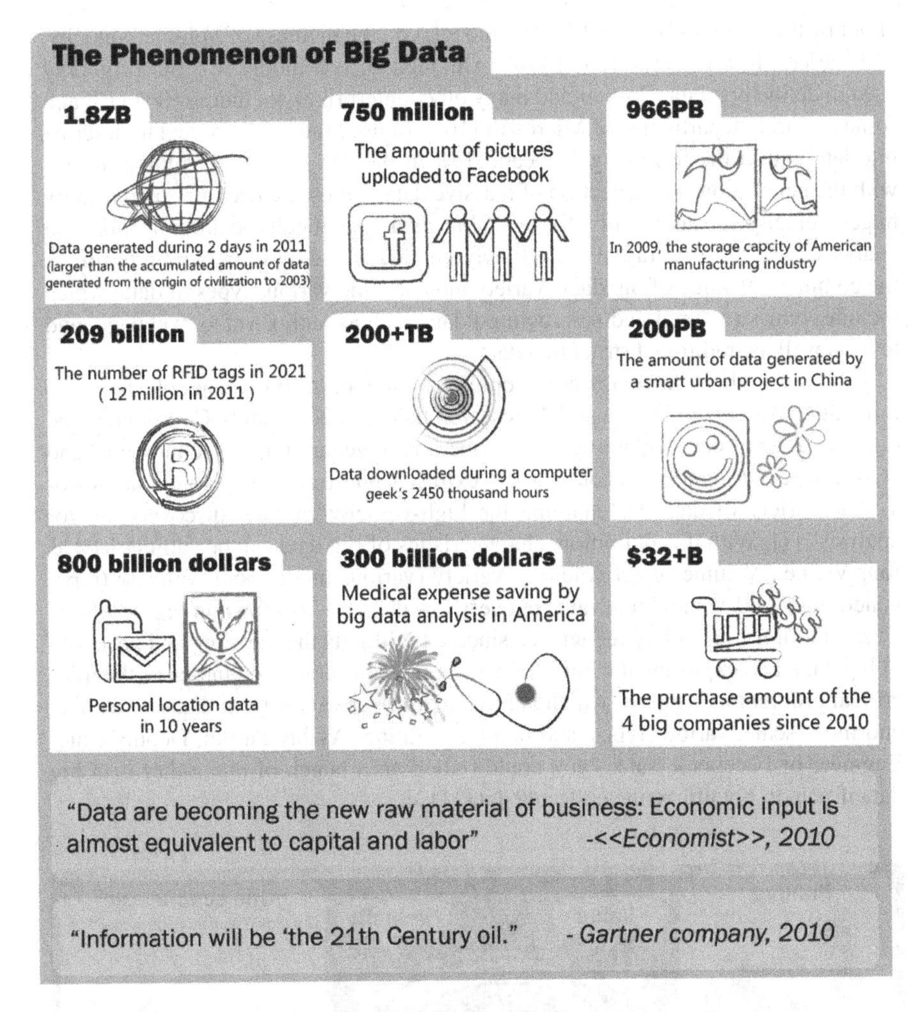

Fig. 1.1 Illustrating the continuously increasing big data

not be acquired, stored, and managed by classic database software. This definition includes two connotations: First, the dataset volumes that conform to the standard of big data are changing, and may grow over time or with technological advances; Second, the dataset volumes that conform to the standard of big data in different applications differ from each other. At present, big data generally range from several TB to several PB [12]. From the definition by McKinsey & Company, it can be seen that the volume of a dataset is not the only criterion for big data. The increasingly growing data scale and its management that could not be handled by traditional database technologies are the next two key features.

As a matter of fact, big data has been defined as early as 2001. Doug Laney, an analyst of META (presently Gartner) defined challenges and opportunities brought about by the increased data with a 3Vs model, i.e., the increase of Volume, Velocity, and Variety, in a research report [14]. Although such a model was not originally used to define big data, Gartner and many other enterprises, including IBM [15] and some research departments of Microsoft [16] still used the "3Vs" model to describe big data within the following 10 years [17]. In the "3Vs" model, Volume means, with the generation and collection of massive data, data scale becomes increasingly huge; Velocity means the timeliness of big data, specifically, data collection and analysis, etc., must be rapidly and timely conducted, so as to maximumly utilize the commercial value of big data; Variety indicates the various types of data, which include semi-structured and unstructured data such as audio, video, webpage, and text, as well as traditional structured data.

However, others have different opinions, including IDC, one of the most influential leaders in big data and its research fields. In 2011, an IDC report defined big data as "big data technologies describe a new generation of technologies and architectures, designed to economically extract value from very large volumes of a wide variety of data, by enabling the high-velocity capture, discovery, and/or analysis" [1]. With this definition, characteristics of big data can be summarized as four Vs, i.e., Volume (great volume), Variety (various modalities), Velocity (rapid generation), and Value (huge value but very low density), as shown in Fig. 1.2. Such 4Vs definition was widely recognized since it highlights the meaning and necessity of big data, i.e., exploring the huge hidden values. This definition indicates the most critical problem in big data, which is how to discover values from datasets with an enormous scale, various types, and rapid generation. As Jay Parikh, Deputy Chief Engineer of Facebook, said, "you could only own a bunch of data other than big data if you do not utilize the collected data" [13].

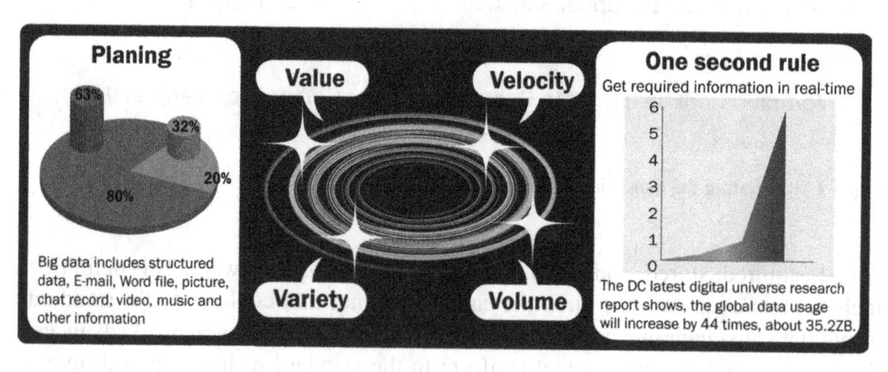

Fig. 1.2 The 4Vs feature of big data

In addition, the US National Institute of Standards and Technology (NIST) defines big data as "Big data shall mean the data of which the data volume, acquisition speed, or data representation limits the capacity of using traditional relational methods to conduct effective analysis or the data which may be effectively processed with important horizontal zoom technologies," which focuses on the technological aspect of big data. It indicates that efficient methods or technologies need to be developed and used to analyze and process big data.

There have been considerable discussions from both industry and academia on the definition of big data [10, 11]. In addition to developing a proper definition, the big data research should also focus on how to extract its value, how to make use of data, and how to transform"a bunch of data" into "big data."

1.3 Big Data Value

McKinsey & Company observed how big data created values after in-depth research on the U.S. healthcare, the EU public sector administration, the U.S. retail, the global manufacturing, and the global personal location data. Through research on the five core industries that represent the global economy, the McKinsey report pointed out that big data may give a full play to the economic function, improve the productivity and competitiveness of enterprises and public sectors, and create huge benefits for consumers. In [12], McKinsey summarized the values that big data could create: if big data could be creatively and effectively utilized to improve efficiency and quality, the potential value of the U.S. medical industry gained through data may surpass USD 300 billion, thus reducing the U.S. healthcare expenditure by over 8 %; retailers that fully utilize big data may improve their profit by more than 60 %; big data may also be utilized to improve the efficiency of government operations, such that the developed economies in Europe could save over EUR 100 billion (which excludes the effect of reduced frauds, errors, and tax difference).

The McKinsey report is regarded as prospective and predictive, while the following facts may validate the values of big data. During the 2009 flu pandemic, Google obtained timely information by analyzing big data, which even provided more valuable information than that provided by disease prevention centers. Nearly all countries required hospitals inform agencies such as disease prevention centers of new type of influenza cases. However, patients usually did not see doctors immediately when they got infected. It also took some time to send information from hospitals to disease prevention centers, and for disease prevention centers to analyze and summarize such information. Therefore, when the public is aware of the pandemic of a new type of influenza, the disease may have already spread for one to two weeks with a serious hysteretic nature. Google found that during the spreading of influenza, entries frequently sought at its search engines would be different from those at ordinary times, and the usage frequencies of the entries were correlated to the influenza spreading in both time and location. Google found 45 search entry groups that were closely relevant to the outbreak of influenza and incorporated them in specific mathematic models to forecast the spreading of influenza and even to predict places where influenza will spread from. The related research results have been published in Nature [18].

In 2008, Microsoft purchased Farecast, a sci-tech venture company in the U.S. Forecast has an airline ticket forecasting system that predicts the trends and rising/dropping ranges of airline ticket prices. The system has been incorporated into the Bing search engine of Microsoft. By 2012, the system has saved nearly USD 50 per ticket per passenger, with the forecast accuracy as high as 75 %.

At present, data has become an important production factor that could be comparable to material assets and human capital. As multimedia, social media, and IoT are fast evolving, enterprises will collect more information, leading to an exponential growth of data volume. Big data will have a huge and increasing potential in creating values for businesses and consumers.

1.4 The Development of Big Data

In late 1970s, the concept of "database machine" emerged, which is a technology specially used for storing and analyzing data. With the increase of data volume, the storage and processing capacity of a single mainframe computer system has become inadequate. In the 1980s, people proposed "share nothing," a parallel database system, to meet the demand of the increasing data volume [19]. The share nothing system architecture is based on the use of cluster and every machine has its own processor, storage, and disk. Teradata system was the first successful commercial parallel database system. Such database became very popular lately. On June 2, 1986, a milestone event occurred, when Teradata delivered the first parallel database system with a storage capacity of 1TB to Kmart to help the large-scale retail company in North America to expand its data warehouse [20]. In late 1990s, the advantages of the parallel database was widely recognized in the database field.

However, many challenges on big data arose. With the development of Internet services, indexes and queried contents were rapidly growing. Therefore, search engine companies had to face the challenges of handling such big data. Google created GFS [21] and MapReduce [22] programming models to cope with the challenges brought about by data management and analysis at the Internet scale. In addition, contents generated by users, sensors, and other ubiquitous data sources also drive the overwhelming data flows, which required a fundamental change on the computing architecture and large-scale data processing mechanism. In January 2007, Jim Gray, a pioneer of database software, called such transformation "The Fourth Paradigm" [23]. He also thought the only way to cope with such a paradigm was to develop a new generation of computing tools to manage, visualize, and analyze massive data. In June 2011, another milestone event occurred, when EMC/IDC published a research report titled *Extracting Values from Chaos* [1], which introduced the concept and potential of big data for the first time. This research report aroused great interest in both industry and academia on big data.

Over the past few years, nearly all major companies, including EMC, Oracle, IBM, Microsoft, Google, Amazon, and Facebook, etc., have started their big data projects. Taking IBM as an example, since 2005, IBM has invested USD 16 billion on 30 acquisitions related to big data. In academia, big data was also under the spotlight. In 2008, Nature published the big data special issue. In 2011, Science also launched a special issue on the key technologies of "data processing" in big data. In 2012, European Research Consortium for Informatics and Mathematics (ERCIM) News published a special issue on big data. In the beginning of 2012, a report titled *Big Data, Big Impact* presented at the Davos Forum in Switzerland, announced that big data has become a new kind of economic assets, just like currency or gold. Gartner, an international research agency, issued *Hype Cycles from 2012 to 2013*, which classified big data computing, social analysis, and stored data analysis into 48 emerging technologies that deserve most attention.

Many national governments such as the U.S. also paid great attention to big data. In March 2012, the Obama Administration announced a USD 200 million investment to launch the *Big Data Research and Development Initiative*, which was a second major scientific and technological development initiative after the *Information Highway Initiative* in 1993. In July 2012, the *Japan's ICT* project issued by Ministry of Internal Affairs and Communications indicated that the big data development should be a national strategy and application technologies should be the focus. In July 2012, the United Nations issued Big Data for Development report, which summarized how governments utilized big data to better serve and protect their people.

1.5 Challenges of Big Data

The sharply increasing data deluge in the big data era brings huge challenges on data acquisition, storage, management and analysis. Traditional data management and analytics systems are based on the relational database management system (RDBMS). However, such RDBMSs only apply to structured data, other than semi-structured or unstructured data. In addition, RDBMSs are increasingly utilizing more and more expensive hardware. It is apparently that the traditional RDBMSs cannot handle the huge volume and heterogeneity of big data. The research community has proposed some solutions from different perspectives. For example, cloud computing is utilized to meet the requirements on infrastructure for big data, e.g., cost efficiency, elasticity, and smooth upgrading/downgrading. For solutions of permanent storage and management of large-scale disordered datasets, distributed file systems [24] and NoSQL [25] databases are good choices. Such programming frameworks have achieved great success in processing clustered tasks, especially for webpage ranking. Various big data applications can be developed based on these innovative technologies or platforms. Moreover, it is non-trivial to deploy the big data analytics systems.

Some literatures [26–28] discuss obstacles to be overcome in the development of big data applications. Some key challenges are listed as follows:

- *Data Representation*: many datasets have certain levels of heterogeneity in type, structure, semantics, organization, granularity, and accessibility. Data representation aims to make data more meaningful for computer analysis and user interpretation. Nevertheless, an improper data representation will reduce the value of the original data and may even obstruct effective data analysis. Efficient data representation shall reflect data structure, class, and type, as well as integrated technologies, so as to enable efficient operations on different datasets.
- *Redundancy Reduction and Data Compression*: generally, there is a high level of redundancy in datasets. Redundancy reduction and data compression is effective to reduce the indirect cost of the entire system on the premise that the potential values of the data are not affected. For example, most data generated by sensor networks are highly redundant, which may be filtered and compressed at orders of magnitude.
- *Data Life Cycle Management*: compared with the relatively slow advances of storage systems, pervasive sensors and computing are generating data at unprecedented rates and scales. We are confronted with a lot of pressing challenges, one of which is that the current storage system could not support such massive data. Generally speaking, values hidden in big data depend on data freshness. Therefore, an importance principle related to the analytical value should be developed to decide which data shall be stored and which data shall be discarded.
- *Analytical Mechanism*: the analytical system of big data shall process masses of heterogeneous data within a limited time. However, traditional RDBMSs are strictly designed with a lack of scalability and expandability, which could not meet the performance requirements. Non-relational databases have shown their unique advantages in the processing of unstructured data and started to become mainstream in big data analysis. Even so, there are still some problems of non-relational databases in their performance and particular applications. We shall find a compromising solution between RDBMSs and non-relational databases. For example, some enterprises have utilized a mixed database architecture that integrates the advantages of both types of database (e.g., Facebook and Taobao). More research is needed on the in-memory database and sample data based on approximate analysis.
- *Data Confidentiality*: most big data service providers or owners at present could not effectively maintain and analyze such huge datasets because of their limited capacity. They must rely on professionals or tools to analyze the data, which increase the potential safety risks. For example, the transactional dataset generally includes a set of complete operating data to drive key business processes. Such data contains details of the lowest granularity and some sensitive information such as credit card numbers. Therefore, analysis of big data may be delivered to a third party for processing only when proper preventive measures are taken to protect the sensitive data, to ensure its safety.

- *Energy Management*: the energy consumption of mainframe computing systems has drawn much attention from both economy and environment perspectives. With the increase of data volume and analytical demands, the processing, storage, and transmission of big data will inevitably consume more and more electric energy. Therefore, system-level power consumption control and management mechanisms shall be established for big data while expandability and accessibility are both ensured.
- *Expendability and Scalability*: the analytical system of big data must support present and future datasets. The analytical algorithm must be able to process increasingly expanding and more complex datasets.
- *Cooperation*: analysis of big data is an interdisciplinary research, which requires experts in different fields cooperate to harvest the potential of big data. A comprehensive big data network architecture must be established to help scientists and engineers in various fields access different kinds of data and fully utilize their expertise, so as to cooperate to complete the analytical objectives.

References

1. John Gantz and David Reinsel. Extracting value from chaos. *IDC iView*, pages 1–12, 2011.
2. Kenneth Cukier. *Data, data everywhere: A special report on managing information*. Economist Newspaper, 2010.
3. Drowning in numbers - digital data will flood the planet- and help us understand it better. http://www.economist.com/blogs/dailychart/2011/11/bigdata-0, 2011.
4. Steve Lohr. The age of big data. *New York Times*, 11, 2012.
5. Noguchi Yuki. Following digital breadcrumbs to big data gold. http://www.npr.org/2011/11/29/142521910/thedigitalbreadcrumbs-that-lead-to-big-data, 2011.
6. Noguchi Yuki. The search for analysts to make sense of big data. http://www.npr.org/2011/11/30/142893065/the-searchforanalysts-to-make-sense-of-big-data, 2011.
7. Big data. http://www.nature.com/news/specials/bigdata/index.html, 2008.
8. Special online collection: Dealing with big data. http://www.sciencemag.org/site/special/data/, 2011.
9. Fact sheet: Big data across the federal government. http://www.whitehouse.gov/sites/default/files/microsites/ostp/big_data_fact_sheet_3_29_2012.pdf, 2012.
10. O. R. Team. *Big Data Now: Current Perspectives from O'Reilly Radar*. O'Reilly Media, 2011.
11. M Grobelnik. Big data tutorial. http://videolectures.net/eswc2012grobelnikbigdata/, 2012.
12. James Manyika, McKinsey Global Institute, Michael Chui, Brad Brown, Jacques Bughin, Richard Dobbs, Charles Roxburgh, and Angela Hung Byers. *Big data: The next frontier for innovation, competition, and productivity*. McKinsey Global Institute, 2011.
13. Viktor Mayer-Schönberger and Kenneth Cukier. *Big Data: A Revolution that Will Transform how We Live, Work, and Think*. Eamon Dolan/Houghton Mifflin Harcourt, 2013.
14. Douglas Laney. 3-d data management: Controlling data volume, velocity and variety. *META Group Research Note, February*, 6, 2001.
15. Paul Zikopoulos, Chris Eaton, et al. *Understanding big data: Analytics for enterprise class hadoop and streaming data*. McGraw-Hill Osborne Media, 2011.
16. Erik Meijer. The world according to linq. *Communications of the ACM*, 54(10):45–51, 2011.
17. Mark Beyer. Gartner says solving 'big data' challenge involves more than just managing volumes of data. *Gartner*. http://www.gartner.com/it/page.jsp, 2011.

18. Jeremy Ginsberg, Matthew H Mohebbi, Rajan S Patel, Lynnette Brammer, Mark S Smolinski, and Larry Brilliant. Detecting influenza epidemics using search engine query data. *Nature*, 457(7232):1012–1014, 2008.
19. David DeWitt and Jim Gray. Parallel database systems: the future of high performance database systems. *Communications of the ACM*, 35(6):85–98, 1992.
20. T Walter. Teradata past, present, and future. UCI ISG Lecture Series on Scalable Data Management.
21. Sanjay Ghemawat, Howard Gobioff, and Shun-Tak Leung. The google file system. In *ACM SIGOPS Operating Systems Review*, volume 37, pages 29–43. ACM, 2003.
22. Jeffrey Dean and Sanjay Ghemawat. Mapreduce: simplified data processing on large clusters. *Communications of the ACM*, 51(1):107–113, 2008.
23. Anthony JG Hey, Stewart Tansley, Kristin Michele Tolle, et al. The fourth paradigm: data-intensive scientific discovery. 2009.
24. John H Howard, Michael L Kazar, Sherri G Menees, David A Nichols, Mahadev Satyanarayanan, Robert N Sidebotham, and Michael J West. Scale and performance in a distributed file system. *ACM Transactions on Computer Systems (TOCS)*, 6(1):51–81, 1988.
25. Rick Cattell. Scalable sql and nosql data stores. *ACM SIGMOD Record*, 39(4):12–27, 2011.
26. Alexandros Labrinidis and HV Jagadish. Challenges and opportunities with big data. *Proceedings of the VLDB Endowment*, 5(12):2032–2033, 2012.
27. Surajit Chaudhuri, Umeshwar Dayal, and Vivek Narasayya. An overview of business intelligence technology. *Communications of the ACM*, 54(8):88–98, 2011.
28. D Agrawal, P Bernstein, E Bertino, S Davidson, U Dayal, M Franklin, J Gehrke, L Haas, A Halevy, J Han, et al. Challenges and opportunities with big data. a community white paper developed by leading researchers across the united states, 2012.

Chapter 2
Related Technologies

Abstract In order to gain a deep understanding of big data, this chapter will introduce several fundamental technologies that are closely related to big data, including cloud computing, Internet of Things (IoT), data center, and Hadoop. For each related technology, a general introduction is first provided highlighting their key features. Then the relationship between the technology and big data is examined in detail.

2.1 Cloud Computing

2.1.1 Cloud Computing Preliminaries

In the big data paradigm, reliable hardware infrastructures is critical to provide reliable storage. The hardware infrastructure includes masses of elastic shared Information and Communications Technology (ICT) resources. Such ICT resources shall be capable of horizontal and vertical expansion and contraction, and dynamic reconfiguration for different applications. Over the years, the advances of cloud computing have been changing the way people acquire and use hardware infrastructure and software services [1].

Cloud Computing is evolved from Distributed Computing, Parallel Computing, and Grid Computing, or a commercial realization of the computer-scientific concept. In a narrow sense, cloud computing means the delivery and use mode of IT infrastructure, i.e., acquiring necessary resources through the Internet on-demand or in an expandable way. In a general sense, cloud computing means the delivery and use mode of services, i.e., acquiring necessary services through the Internet on-demand or in an expandable way. Such service may related to software and the Internet, or others. In short, it refers to the case that users access a server through the network in a remote location and then use some services provided by the server.

M. Chen et al., *Big Data: Related Technologies, Challenges and Future Prospects*,
SpringerBriefs in Computer Science, DOI 10.1007/978-3-319-06245-7_2,

This concept mainly evolves from some mixed concepts such as virtualized public computing and infrastructure. The key components of cloud computing is illustrated in Fig. 2.1.

Services provided by cloud computing can be described by three service models and three deployment models. Such a combination has many important features, including self-service as required, wide network access, resource pool, rapidity, elasticity, and service management, thus meeting the requirements of many applications. Therefore, cloud computing will be instrumental for big data analysis and applications.

2.1.2 Relationship Between Cloud Computing and Big Data

Cloud computing is closely related to big data. The key components of cloud computing are shown in Fig. 2.1. Big data is the object of the computation operation and stresses the storage capacity and computing capacity of a cloud server. The main objective of cloud computing is to use huge computing resources and computing capacities under concentrated management, so as to provide applications with resource sharing at a granularity and provide big data applications with computing capacity. The development of cloud computing provides solutions for the storage and processing of big data. On the other hand, the emergence of big data also accelerates the development of cloud computing. The distributed storage technology based on cloud computing allows effective management of big data; the parallel computing capacity by virtue of cloud computing can improve the efficiency of acquiring and analyzing big data.

Even though there are many overlapped concepts and technologies in cloud computing and big data, they differ in the following two major aspects. First, the concepts are different in the sense that cloud computing transforms the IT architecture while big data influences business decision-making, while big data depends on cloud computing as the fundamental infrastructure for smooth operation.

Second, big data and cloud computing have different target customers. Cloud computing is a technology and product targeting Chief Information Officers (CIO) as an advanced IT solution. Big data is a product targeting Chief Executive Officers (CEO) focusing on business operations. Since the decision makers may directly feel the pressure from market competition, they must defeat business opponents in more competitive ways. With the advances of big data and cloud computing, these two technologies are certainly and increasingly entwine with each other. Cloud computing, with functions similar to those of computers and operating systems, provides system-level resources; big data operates in the upper level supported by cloud computing and provides functions similar to those of database and efficient data processing capacity. As Kissinger, President of EMC, said, the application of big data must be based on cloud computing.

The evolution of big data was driven by the rapid growth of application demands and cloud computing developed from virtualization technologies. Therefore, cloud

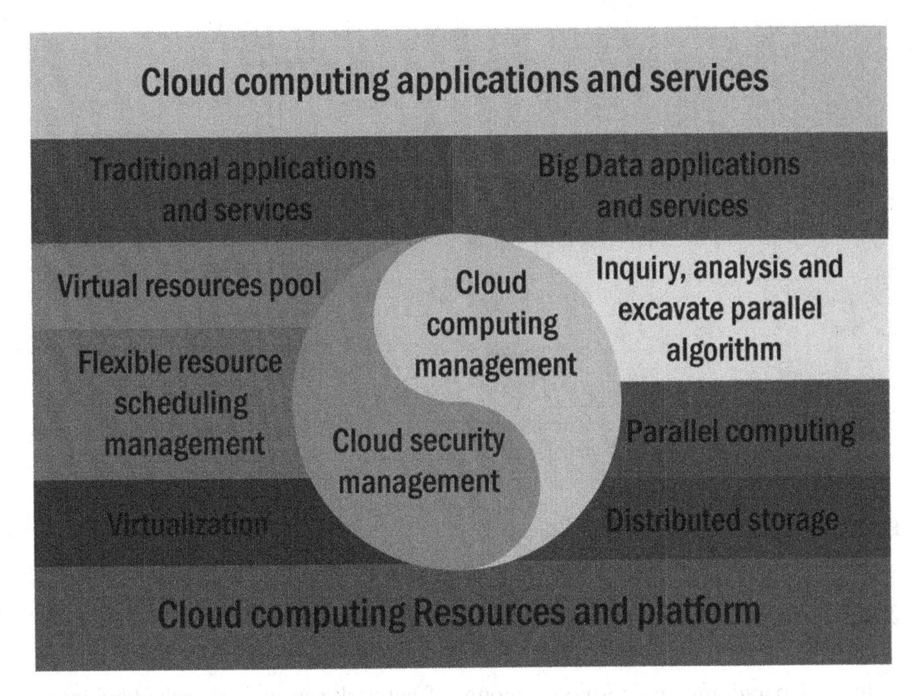

Fig. 2.1 Key components of cloud computing

computing not only provides computation and processing for big data, but also itself is a service mode. To a certain extent, the advances of cloud computing also promote the development of big data, both of which supplement each other.

2.2 IoT

2.2.1 *IoT Preliminaries*

The basic idea of IoT is to connect different objects in the real world, such as RFID, bar code readers, sensors, and mobile phones, etc., to realize information exchange and to make them cooperate with each other to complete a common task. The IoT architecture is illustrated in Fig. 2.2. IoT is deemed as the extension of the Internet and is an important part of the future Internet. IoT is mainly characterized with that it accesses every object in the physical world such that the objects can be addressed, controlled, and communicated with.

Compared with the Internet, IoT has the following main features [2].

- Various terminal equipments
- Automatic data acquisition
- Intelligent terminals

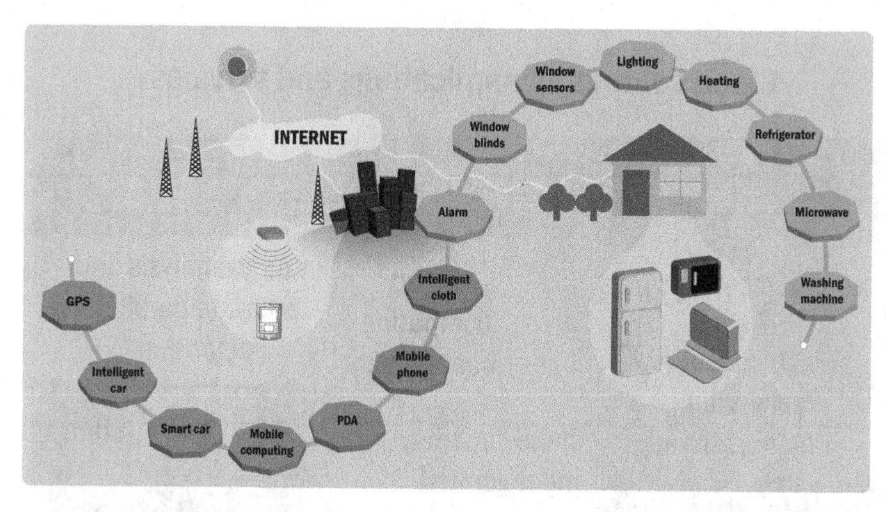

Fig. 2.2 Illustration of the IoT architecture

2.2.2 Relationship Between IoT and Big Data

In the IoT paradigm, an enormous amount of network sensors are embedded into devices in the real world. Such sensors deployed in different fields may collect various kinds of data, such as environmental data, geographical data, astronomical data, and logistic data. Mobile equipments, transportation facilities, public facilities, and home appliances could all be data acquisition equipment in IoT.

The big data generated by IoT has different characteristics compared with general big data because of the different types of data collected, of which the most classical characteristics include heterogeneity, variety, unstructured feature, noise, and rapid growth. Although the current IoT data is not the dominant part of big data, by 2030, the quantity of sensors will reach one trillion and then the IoT data could be the most important part of big data, according to the forecast of HP. A report from Intel pointed out that big data in IoT has three features that conform to the big data paradigm: (a) abundant terminals generating masses of data; (b) data generated by IoT is usually semi-structured or unstructured; (c) data of IoT is useful only when it is analyzed.

At present, the data processing capacity of IoT has fallen behind the collected data and it is extremely urgent to accelerate the introduction of big data technologies to catch up with the development of IoT. Many operators of IoT realize the importance of big data since the success of IoT is hinged upon the effective integration of big data and cloud computing. The widespread deployment of IoT will also bring many cities into the big data era.

There is a compelling need to adopt big data for IoT applications, while the development of big data is already legged behind. It has been widely recognized that these two technologies are inter-dependent and should be jointly developed.

On one hand, the widespread deployment of IoT drives the high growth of data both in quantity and category, thus providing the opportunity for the application and development of big data. On the other hand, the application of big data technology to IoT also accelerates the research advances and business models of IoT.

2.3 Data Center

In the big data paradigm, a data center is not only an organization for concentrated storage of data, but also undertakes more responsibilities, such as acquiring data, managing data, organizing data, and leveraging the data values and functions. Data centers are mainly concerned with "data" other than "center." A data center has masses of data and organizes and manages data according to its core objective and development path, which is more valuable than owning a good site and resource. The emergence of big data brings about abundant development opportunities and great challenges to data centers.

- Big data requires data center provide powerful backstage support. The big data paradigm has more stringent requirements on storage capacity and processing capacity, as well as network transmission capacity. Enterprises must take the development of data centers into consideration to improve the capacity of rapidly and effectively processing of big data under limited price/performance ratio. The data center shall provide the infrastructure with a large number of nodes, build a high-speed internal network, effectively dissipate heat, and effective backup data. Only when a highly energy-efficient, stable, safe, expandable, and redundant data center is built, the normal operation of big data applications may be ensured.
- The growth of big data applications accelerates the revolution and innovation of data centers. Many big data applications have developed their unique architectures and directly promote the development of storage, network, and computing technologies related to data center. With the continued growth of structured and unstructured data, and the variety of sources of analytical data, the data processing and computing capacities of the data center shall be greatly enhanced. In addition, as the scale of data center is increasingly expanding, it is also an important issue on how to reduce the operational cost for the development of data centers.
- Big data endows more functions to data centers. In the big data paradigm, a data center shall not only be concerned with hardware facilities but also strengthen soft capacities, i.e., the capacities of acquisition, processing, organization, analysis, and application of big data. The data center may help business personnel analyze the existing data, discover problems in business operation, and develop solutions from big data.

Big data is an emerging paradigm, which will promote the explosive growth of the infrastructure and related software of data center. The physical data center network is the core for supporting big data, but, at present, is the key infrastructure that is most urgently in need [3].

2.4 Hadoop

2.4.1 *Hadoop Preliminaries*

Hadoop is a technology closely related to big data, which forms a powerful big data systematic solution through data storage, data processing, system management, and integration of other modules. Such technology has become indispensible to cope with the challenges of big data [4]. Hadoop is a set of large-scale software infrastructures for Internet applications similar to Google's FileSystem and MapReduce. Hadoop was developed by Nutch, an open-source project of Apache, with the initial design completed by Doug Cutting and Mike Cafarella. In 2006, Hadoop became an independent open-source project of Apache, which is widely deployed by Yahoo, Facebook, and other Internet enterprises. At present, the biggest Hadoop cluster operated by Yahoo has 4,000 sets of nodes used for data processing and analysis, including Yahoo's advertisements, financial data, and user logs.

Hadoop consists of two parts: HDFS (Hadoop Distributed File System) and MR framework (MapReduce Framework). HDFS is the data storage source of MR, which is a distributed file system running on commercial hardware and designed in reference to Google's DFS. HDFS is the basis for main data storage of Hadoop applications, which distributes files in data blocks of 64MB and stores such data blocks in different nodes of a cluster, so as to enable parallel computing for MR. An HDFS cluster includes a single NameNode for managing the metadata of the file system and DataNodes for storing actual data. A file is divided into one or multiple blocks and such blocks are stored in DataNodes. Copies of blocks are distributed to different DataNodes to prevent data loss. Apache HBase is a column-oriented storage, which imitates GooglesBigtable. Therefore, functions of Apache HBase are similar to those of BigTable as described in Part VI of HDFS. HBase may be taken as an input and output server of the MR task of Hadoop, and be accessed through Java API, REST, Avor, or Thrift APIs.

MR was developed similar to MapReduce of Google. The MR framework consists of one JobTracker node and multiple TaskTracker nodes. The JobTracker node is used for task distribution and task scheduling; TaskTracker nodes are used to receive Map or Reduce tasks distributed from JobTracker node and execute such tasks and feed task status back to the JobTracker node. MR framework and HDFS run in the same node set, so as to schedule tasks on nodes presented with data. Pig Latin is a high-level declarative language, which can describe the big data aggregation and analysis tasks in MR programming. Hive supports queries expressed by declarative similar to HiveQL and SQL. Hive introduces the concept of RDBMSs and SQL subset people are familiar with to Hadoop.

Apart from the aforementioned core parts, other modules related to Hadoop may also provide some supplementary functions required in the value chain of big data. Zookeeper and Chukwa are used to manage and monitor distributed applications run in Hadoop. It is worth noting that Zookeeper is the central service to maintain configuration and naming, provide distributed synchronization, and provide grouped services. Chukwa is responsible for monitoring system status and can display, monitor, and analyze collected data. Sqoop allows data to be conveniently passed between the structured data storage and Hadoop. Mahout is a data mining base executed on Hadoop using MapReduce. The base includes core algorithms of collaborative filtering used for clustering and sorting, and is based on batch processing.

Benefited from the huge success of the distributed file system of Google and the computational model of MapReduce for processing massive data, Hadoop, its clone, attracts more and more attentions. Hadoop is closely related to big data as nearly all leading enterprises of big data have commercial big data solutions based on Hadoop. Hadoop is becoming the corner stone of big data. Apache Hadoop is an open-source software framework. Hadoop realizes the distributed processing of massive data in the large-scale commercial server cluster, other than relying on expensive exclusive hardware and various systems to store and process data.

Hadoop has many advantages, but the following aspects are especially relevant to the management and analysis of big data:

- *Expandability*: Hadoop allows the expansion or shrinkage of hardware infrastructure without changing data format. The system will automatically re-distribute data and computing tasks will be adapted to hardware changes.
- *High Cost Efficiency*: Hadoop applies large-scale parallel computing to commercial servers, which greatly reduces the cost per TB required for storage capacity. The large-scale computing also enables it to accommodate the continually growing data volume.
- *Strong Flexibility*: Hadoop may handle many kinds of data from various sources. In addition, data from many sources can be synthesized in Hadoop for further analysis. Therefore, it can cope with many kinds of challenges brought by big data.
- *High Fault-Tolerance*: it is common that data loss and miscalculation occur during the analysis of big data, but Hadoop can recover data and correct computing errors caused by node failures or network congestion.

2.4.2 Relationship between Hadoop and Big Data

Presently, Hadoop is widely used in big data applications in the industry, e.g., spam filtering, network searching, clickstream analysis, and social recommendation. In addition, considerable academic research is now based on Hadoop. Some representative cases are given below. As declared in June 2012, Yahoo runs Hadoop in 42,000 servers at four data centers to support its products and services, e.g., searching and spam filtering, etc. At present, the biggest Hadoop cluster has 4,000 nodes, but the number of nodes will be increased to 10,000 with the release of Hadoop 2.0. In the same month, Facebook announced that their Hadoop cluster can process 100 PB data, which grew by 0.5 PB per day as in November 2012. Some well-known agencies that use Hadoop to conduct distributed computation are listed in [5]. In addition, many companies provide Hadoop commercial execution and support, including Cloudera, IBM, MapR, EMC, and Oracle.

Among modern industrial machinery and systems, sensors are widely deployed to collect information for environment monitoring and failure forecasting, etc. Bahga and others in [6] proposed a framework for data organization and cloud computing infrastructure, termed CloudView. CloudView uses mixed architectures, local nodes, and remote clusters based on Hadoop to analyze machine-generated data. Local nodes are used for the forecast of real-time failures; clusters based on Hadoop are used for complex offline analysis, e.g., case-driven data analysis.

The exponential growth of the genome data and the sharp drop of sequencing cost transform bio-science and bio-medicine to data-driven science. Gunarathne et al. in [7] utilized cloud computing infrastructures, Amazon AWS, Microsoft Azune, and data processing framework based on MapReduce, Hadoop, and Microsoft DryadLINQ to run two parallel bio-medicine applications: (a) assembly of genome segments; (b) dimension reduction in the analysis of chemical structure. In the subsequent application, the 166-D datasets used include 26,000,000 data points. The authors compared the performance of all the frameworks in terms of efficiency, cost, and availability. According to the study, the authors concluded that the loose coupling will be increasingly applied to research on electron cloud, and the parallel programming technology (i.e., MapReduce) framework may provide users an interface with more convenient services and reduce unnecessary costs.

References

1. Giuseppe DeCandia, Deniz Hastorun, Madan Jampani, Gunavardhan Kakulapati, Avinash Lakshman, Alex Pilchin, Swaminathan Sivasubramanian, Peter Vosshall, and Werner Vogels. Dynamo: amazon's highly available key-value store. In *SOSP*, volume 7, pages 205–220, 2007.
2. Luigi Atzori, Antonio Iera, and Giacomo Morabito. The internet of things: A survey. *Computer Networks*, 54(15):2787–2805, 2010.
3. Yantao Sun, Min Chen, Bin Liu, and Shiwen Mao. Far: A fault-avoidant routing method for data center networks with regular topology. In *Proceedings of ACM/IEEE Symposium on Architectures for Networking and Communications Systems (ANCS'13)*. ACM, 2013.
4. Tom White. *Hadoop: the definitive guide*. O'Reilly, 2012.
5. Wiki. Applications and organizations using hadoop. http://wiki.apache.org/hadoop/PoweredBy, 2013.
6. Arshdeep Bahga and Vijay K Madisetti. Analyzing massive machine maintenance data in a computing cloud. *Parallel and Distributed Systems, IEEE Transactions on*, 23(10):1831–1843, 2012.
7. Thilina Gunarathne, Tak-Lon Wu, Jong Youl Choi, Seung-Hee Bae, and Judy Qiu. Cloud computing paradigms for pleasingly parallel biomedical applications. *Concurrency and Computation: Practice and Experience*, 23(17):2338–2354, 2011.

Chapter 3
Big Data Generation and Acquisition

Abstract We have introduced several key technologies related to big data, i.e., cloud computing, IoT, data center, and Hadoop. Next, we will focus on the value chain of big data, which can be generally divided into four phases: data generation, data acquisition, data storage, and data analysis. If we take data as a raw material, data generation and data acquisition are exploitation process, data storage is a storage process, and data analysis is a production process that utilizes the raw material to create new value.

3.1 Big Data Generation

Data generation is the first step of big data. Specifically, it is large-scale, highly diverse, and complex datasets generated through longitudinal and distributed data sources. Such data sources include sensors, videos, click streams, and/or all other available data sources. At present, main sources of big data are the operation and trading information in enterprises, logistic and sensing information in the IoT, human interaction information and position information in the Internet world, and data generated in scientific research, etc. The information far surpasses the capacities of IT architectures and infrastructures of existing enterprises, while its real-time requirement also greatly stresses the existing computing capacity.

3.1.1 Enterprise Data

In 2013, IBM issued a reported titled "Analytics: The Real-world Use of Big Data," which indicates that the internal data of enterprises are the main sources of big data. The internal data of enterprises mainly consists of online trading data and online analysis data, most of which are historically static data and are managed by RDBMSs in a structured manner. In addition, production data, inventory data, sales

M. Chen et al., *Big Data: Related Technologies, Challenges and Future Prospects*, SpringerBriefs in Computer Science, DOI 10.1007/978-3-319-06245-7_3,

data, and financial data, etc., also constitute enterprise internal data, which aims to capture informationized and data-driven activities in enterprises, so as to record all activities of enterprises in the form of internal data.

Over the past decades, IT and digital data have contributed a lot to improve the profitability of business departments. It is estimated that the business data volume of all companies in the world may double every 1.2 years [1], in which, the business turnover through the Internet, enterprises to enterprises, and enterprises to consumers per day will reach USD 450 billion [2]. The continuously increasing business data volume requires more effective real-time analysis so as to fully harvest its potential. For example, Amazon processes millions of terminal operations and more than 500,000 queries from third-party sellers per day [3]. Walmart processes one million customer trades per hour and such trading data are imported into a database with a capacity of over 2.5PB [4]. Akamai analyzes 75 million events per day for its target advertisements [5].

3.1.2 *IoT Data*

As discussed, IoT is an important source of big data. Among smart cities constructed based on IoT, big data may come from industry, agriculture, traffic and transportation, medical care, public departments, and households, etc., as shown in Fig. 3.1.

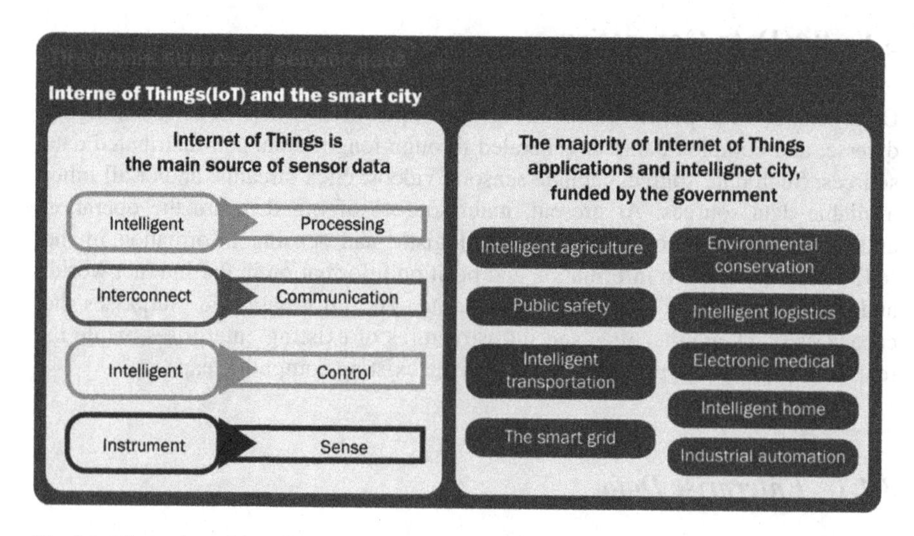

Fig. 3.1 Illustration of the prime source of sensory data

According to the processes of data acquisition and transmission in IoT, its network architecture may be divided into three layers: the sensing layer, the network layer, and the application layer. The sensing layer is responsible for data acquisition and mainly consists of sensor networks. The network layer is responsible for information transmission and processing, where close transmission may rely on sensor networks, and remote transmission shall depend on the Internet. Finally, the application layer support specific applications of IoT.

According to the characteristics of IoT, the data generated from IoT has the following features:

- *Large-Scale Data*: in IoT, masses of data acquisition equipments are distributedly deployed, which may acquire simple numeric data (e.g., location) or complex multimedia data (e.g., surveillance video). In order to meet the demands of analysis and processing, not only the currently acquired data, but also the historical data within a certain time frame should be stored. Therefore, data generated by IoT are characterized by large scales.
- *Heterogeneity*: because of the variety data acquisition devices, the acquired data is also different and such data features heterogeneity.
- *Strong Time and Space Correlation*: in IoT, every data acquisition device are placed at a specific geographic location and every piece of data has a time stamp. The time and space correlations are important properties of data from IoT. During data analysis and processing, time and space are also important dimensions for statistical analysis.
- *Effective Data Accounts for Only a Small Portion of the Big Data*: a great quantity of noises may occur during the acquisition and transmission of data in IoT. Among datasets acquired by acquisition devices, only a small amount of abnormal data is valuable. For example, during the acquisition of traffic video, the few video frames that capture the violation of traffic regulations and traffic accidents are more valuable than those only capturing the normal flow of traffic.

3.1.3 Internet Data

Internet data consists of searching entries, Internet forum posts, chatting records, and microblog messages, among others, which have similar features, such as high value and low density. Such Internet data may be valueless individually, but through exploitation of accumulated big data, useful information such as habits and hobbies of users can be identified, and it is even possible to forecast users' behavior and emotional moods.

3.1.4 Bio-medical Data

As a series of high-throughput bio-measurement technologies are innovatively developed in the beginning of the twenty-first century, the frontier research in the bio-medicine field also enters the era of big data. By constructing smart, efficient, and accurate analytical models and theoretical systems for bio-medicine applications, the essential governing mechanism behind complex biological phenomena may be revealed. Not only the future development of bio-medicine can be determined, but also the leading roles can be assumed in the development of a series of important strategic industries related to the national economy, people's livelihood, and national security, with important applications such as medical care, new drug R&D, and grain production (e.g., transgenic crops).

The completion of HGP (Human Genome Project) and the continued development of sequencing technology also lead to widespread applications of big data in the field. The masses of data generated by gene sequencing go through specialized analysis according to different application demands, to combine it with the clinical gene diagnosis and provide valuable information for early diagnosis and personalized treatment of disease. One sequencing of human gene may generate 100–600 GB raw data. In the China National Genebank in Shenzhen, there are 1.3 million samples including 1.15 million human samples and 150,000 animal, plant, and microorganism samples. By the end of 2013, 10 million traceable biological samples will be stored, and by the end of 2015, this figure will reach 30 million. It is predictable that, with the development of bio-medicine technologies, gene sequencing will become faster and more convenient, and thus making big data of bio-medicine continuously grow beyond all doubt.

In addition, data generated from clinical medical care and medical R&D also rise quickly. For example, the University of Pittsburgh Medical Center (UPMC) has stored 2TB such data. Explorys, an American company, provides platforms to collocate clinical data, operation and maintenance data, and financial data. At present, about 13 million people's information have been collocated, with 44 articles of data at the scale of about 60TB, which will reach 70TB in 2013. Practice Fusion, another American company, manages electronic medical records of about 200,000 patients.

Apart from such small and medium-sized enterprises, other well-known IT companies, such as Google, Microsoft, and IBM have invested extensively in the research and computational analysis of methods related to high-throughput biological big data, for shares in the huge market as known as the "Next Internet." IBM forecasts, in the 2013 Strategy Conference, that with the sharp increase of medical images and electronic medical records, medical professionals may utilize big data to extract useful clinical information from masses of data to obtain a medical history and forecast treatment effects, thus improving patient care and reduce cost. It is anticipated that, by 2015, the average data volume of every hospital will increase from 167TB to 665TB.

3.1.5 Data Generation from Other Fields

As scientific applications are increasing, the scale of datasets is gradually expanding, and the development of some disciplines greatly relies on the analysis of masses of data. Here, we examine several such applications. Although being in different scientific fields, the applications have similar and increasing demand on data analysis. The first example is related to computational biology. GenBank is a nucleotide sequence database maintained by the U.S. National Bio-Technology Innovation Center. Data in this database may double every 10 months. By August 2009, Genbank has more than 250 billion bases from 150,000 different organisms [6]. The second example is related to astronomy. Sloan Digital Sky Survey (SDSS), the biggest sky survey project in astronomy, has recorded 25TB data from 1998 to 2008. As the resolution of the telescope is improved, by 2004, the data volume generated per night will surpass 20TB. The last application is related to high-energy physics. In the beginning of 2008, the Atlas experiment of Large Hadron Collider (LHC) of European Organization for Nuclear Research generates raw data at 2PB/s and stores about 10TB processed data per year.

In addition, pervasive sensing and computing among nature, commercial, Internet, government, and social environments are generating heterogeneous data with unprecedented complexity. These datasets have their unique data characteristics in scale, time dimension, and data category. For example, mobile data were recorded with respect to positions, movement, approximation degrees, communications, multimedia, use of applications, and audio environment. According to the application environment and requirements, such datasets can be classified into different categories, so as to select the proper and feasible solutions for big data.

3.2 Big Data Acquisition

As the second phase of the big data system, big data acquisition includes data collection, data transmission, and data pre-processing. During big data acquisition, once the raw data is collected, an efficient transmission mechanism should be used to send it to a proper storage management system to support different analytical applications. The collected datasets may sometimes include much redundant or useless data, which unnecessarily increases storage space and affects the subsequent data analysis. For example, high redundancy is very common among datasets collected by sensors for environment monitoring. Data compression techniques can be applied to reduce the redundancy. Therefore, data pre-processing operations are indispensable to ensure efficient data storage and exploitation.

3.2.1 Data Collection

Data collection is to utilize special data collection techniques to acquire raw data from a specific data generation environment. Four common data collection methods are shown as follows.

- *Log Files*: As one widely used data collection method, log files are record files automatically generated by the data source system, so as to record activities in designated file formats for subsequent analysis. Log files are typically used in nearly all digital devices. For example, web servers record in log files number of clicks, click rates, visits, and other property records of web users [7]. To capture activities of users at the web sites, web servers mainly include the following three log file formats: public log file format (NCSA), expanded log format (W3C), and IIS log format (Microsoft). All the three types of log files are in the ASCII text format. Databases other than text files may sometimes be used to store log information to improve the query efficiency of the massive log store [8,9]. There are also some other log files based on data collection, including stock indicators in financial applications and determination of operating states in network monitoring and traffic management.
- *Sensors*: Sensors are common in daily life to measure physical quantities and transform physical quantities into readable digital signals for subsequent processing (and storage). Sensory data may be classified as sound wave, voice, vibration, automobile, chemical, current, weather, pressure, temperature, etc. Sensed information is transferred to a data collection point through wired or wireless networks. For applications that may be easily deployed and managed, e.g., video surveillance system [10], the wired sensor network is a convenient solution to acquire related information. Sometimes the accurate position of a specific phenomenon is unknown, and sometimes the monitored environment does not have the energy or communication infrastructures. Then wireless communication must be used to enable data transmission among sensor nodes under limited energy and communication capability. In recent years, WSNs have received considerable interest and have been applied to many applications, such as environmental research [11, 12], water quality monitoring [13], civil engineering [14, 15], and wildlife habit monitoring [16]. A WSN generally consists of a large number of geographically distributed sensor nodes, each being a micro device powered by battery. Such sensors are deployed at designated positions as required by the application to collect remote sensing data. Once the sensors are deployed, the base station will send control information for network configuration/management or data collection to sensor nodes. Based on such control information, the sensory data is assembled in different sensor nodes and sent back to the base station for further processing. Interested readers are referred to [17] for more detailed discussions.
- *Methods for Acquiring Network Data*: At present, network data acquisition is accomplished using a combination of web crawler, word segmentation system, task system, and index system, etc. Web crawler is a program used by search engines for downloading and storing web pages [18]. Generally speaking, web crawler starts from the uniform resource locator (URL) of an initial web page to access other linked web pages, during which it stores and sequences all the retrieved URLs. Web crawler acquires a URL in the order of precedence through a URL queue and then downloads web pages, and identifies all URLs in the downloaded web pages, and extracts new URLs to be put in the queue. This

process is repeated until the web crawler is stopped. Data acquisition through a web crawler is widely applied in applications based on web pages, such as search engines or web caching. Traditional web page extraction technologies feature multiple efficient solutions and considerable research has been done in this field. As more advanced web page applications are emerging, some extraction strategies are proposed in [19] to cope with rich Internet applications.

The current network data acquisition technologies mainly include traditional Libpcap-based packet capture technology, zero-copy packet capture technology, as well as some specialized network monitoring software such as Wireshark, SmartSniff, and WinNetCap.

- *Libpcap-Based Packet Capture Technology*: Libpcap (packet capture library) is a widely used network data packet capture function library. It is a general tool that does not depend on any specific system and is mainly used to capture data in the data link layer. It features simplicity, easy-to-use, and portability, but has a relatively low efficiency. Therefore, under a high-speed network environment, considerable packet losses may occur when Libpcap is used.
- *Zero-Copy Packet Capture Technology*: The so-called zero-copy (ZC) means that no copies between any internal memories occur during packet receiving and sending at a node. In sending, the data packets directly start from the user buffer of applications, pass through the network interfaces, and arrive at an external network. In receiving, the network interfaces directly send data packets to the user buffer. The basic idea of zero-copy is to reduce data copy times, reduce system calls, and reduce CPU load while datagrams are passed from network equipments to user program space. The zero-copy technology first utilizes direct memory access (DMA) technology to directly transmit network datagrams to an address space pre-allocated by the system kernel, so as to avoid the participation of CPU. In the meanwhile, it maps the internal memory of the datagrams in the system kernel to the that of the detection program, or builds a cache region in the user space and maps it to the kernel space. Then the detection program directly accesses the internal memory, so as to reduce internal memory copy from system kernel to user space and reduce the amount of system calls.
- *Mobile Equipments*: At present, mobile devices are more widely used. As mobile device functions become increasingly stronger, they feature more complex and multiple means of data acquisition as well as more variety of data. Mobile devices may acquire geographical location information through positioning systems; acquire audio information through microphones; acquire pictures, videos, streetscapes, two-dimensional barcodes, and other multimedia information through cameras; acquire user gestures and other body language information through touch screens and gravity sensors. Over the years, wireless operators have improved the service level of the mobile Internet by acquiring and analyzing such information. For example, iPhone itself is a "mobile spy." It may collect wireless data and geographical location information, and then send such information back to Apple Inc. for processing, of which the user may not be

aware. Apart from Apple, smart phone operating systems such as Android of Google and Windows Phone of Microsoft can also collect information in the similar manner.

In addition to the aforementioned three data acquisition methods of main data sources, there are many other data collect methods or systems. For example, in scientific experiments, many special tools can be used to collect experimental data, such as magnetic spectrometers and radio telescopes. We may classify data collection methods from different perspectives. From the perspective of data sources, data collection methods can be classified into two categories: collection methods recording through data sources and collection methods recording through other auxiliary tools.

3.2.2 Data Transportation

Upon the completion of raw data collection, data will be transferred to a data storage infrastructure for processing and analysis. As discussed in Sect. 2.3, big data is mainly stored in a data center. The data layout should be adjusted to improve computing efficiency or facilitate hardware maintenance. In other words, internal data transmission may occur in the data center. Therefore, data transmission consists of two phases: Inter-DCN transmissions and Intra-DCN transmissions.

Inter-DCN transmissions are from data source to data center, which is generally achieved with the existing physical network infrastructure. Because of the rapid growth of traffic demands, the physical network infrastructure in most regions around the world are constituted by high-volume, high-rate, and cost-effective optic fiber transmission systems. Over the past 20 years, advanced management equipment and technologies have been developed, such as IP-based wavelength division multiplexing (WDM) network architecture, to conduct smart control and management of optical fiber networks [20, 21]. WDM is a technology that multiplexes multiple optical carrier signals with different wave lengths and couples them to the same optical fiber of the optical link. In such technology, lasers with different wave lengths carry different signals. By far, the backbone network have been deployed with WDM optical transmission systems with single channel rate of 40 Gb/s. At present, 100 Gb/s commercial interface are available and 100 Gb/s systems (or TB/s systems) will be available in the near future [22].

However, traditional optical transmission technologies are limited by the bandwidth of the electronic bottleneck [23]. Recently, orthogonal frequency-division multiplexing (OFDM), initially designed for wireless systems, is regarded as one of the main candidate technologies for future high-speed optical transmission. OFDM is a multi-carrier parallel transmission technology. It segments a high-speed data flow to transform it into low-speed sub-data-flows to be transmitted over multiple orthogonal sub-carriers [24]. Compared with fixed channel spacing of WDM, OFDM allows sub-channel frequency spectrums to overlap with each other [25]. Therefore, it is a flexible and efficient optical networking technology.

Intra-DCN transmissions are the data communication flows within data centers. Intra-DCN transmissions depend on the communication mechanism within the data center (i.e., on physical connection plates, chips, internal memories of data servers, network architectures of data centers, and communication protocols). A data center consists of multiple integrated server racks interconnected with its internal connection networks. Nowadays, the internal connection networks of most data centers are fat-tree, two-layer or three-layer structures based on multi-commodity network flows [23, 26]. In the two-layer topological structure, the racks are connected by 1 Gbps top rack switches (TOR) and then such top rack switches are connected with 10 Gbps aggregation switches in the topological structure. The three-layer topological structure is a structure augmented with one layer on the top of the two-layer topological structure and such layer is constituted by 10 or 100 Gbps core switches to connect aggregation switches in the topological structure. There are also other topological structures which aim to improve the data center networks [27–30].

Because of the inadequacy of electronic packet switches, it is difficult to increase communication bandwidths while keeps energy consumption is low. Over the years, due to the huge success achieved by optical technologies, the optical interconnection among the networks in data centers has drawn great interest. Optical interconnection is a high-throughput, low-delay, and low-energy-consumption solution. At present, optical technologies are only used for point-to-point links in data centers. Such optical links provide connection for the switches using the low-cost multi-mode fiber (MMF) with 10 Gbps data rate. Optical interconnection (switching in the optical domain) of networks in data centers is a feasible solution, which can provide Tbps-level transmission bandwidth with low energy consumption.

Recently, many optical interconnection plans are proposed for data center networks [31]. Some plans add optical paths to upgrade the existing networks, and other plans completely replace the current switches [31–36]. As a strengthening technology, Zhou et al. in [37] adopt wireless links in the 60 GHz frequency band to strengthen wired links. Network virtualization should also be considered to improve the efficiency and utilization of data center networks.

3.2.3 *Data Pre-processing*

Because of the wide variety of data sources, the collected datasets vary with respect to noise, redundancy, and consistency, etc., and it is undoubtedly a waste to store meaningless data. In addition, some analytical methods have stringent requirements on data quality. Therefore, data should be pre-processed under many circumstances to integrate the data from different sources, so as to enable effective data analysis. Pre-processing data not only reduces storage expense, but also improves analysis accuracy. Some relational data pre-processing techniques are discussed in the following.

3.2.3.1 Integration

Data integration is the cornerstone of modern commercial informatics, which involves the combination of data from different sources and provides users with a uniform view of data [38]. This is a mature research field for traditional database. Historically, two methods have been widely recognized: data warehouse and data federation. Data warehousing includes a process named ETL (Extract, Transform and Load). Extraction involves connecting source systems, selecting, collecting, analyzing, and processing necessary data. Transformation is the execution of a series of rules to transform the extracted data into standard formats. Loading means importing extracted and transformed data into the target storage infrastructure. Loading is the most complex procedure among the three, which includes operations such as transformation, copy, clearing, standardization, screening, and data organization. A virtual database can be built to query and aggregate data from different data sources, but such database does not contain data. On the contrary, it includes information or metadata related to actual data and its positions. Such two "storage-reading" approaches do not satisfy the high performance requirements of data flows or search programs and applications. Compared with queries, data in such two approaches is more dynamic and must be processed during data transmission. Generally, data integration methods are accompanied with flow processing engines and search engines [39, 40].

3.2.3.2 Cleaning

Data cleaning is a process to identify inaccurate, incomplete, or unreasonable data, and then modify or delete such data to improve data quality. Generally, data cleaning includes five complementary procedures [41]: defining and determining error types, searching and identifying errors, correcting errors, documenting error examples and error types, and modifying data entry procedures to reduce future errors. During cleaning, data formats, completeness, rationality, and restriction shall be inspected. Data cleaning is of vital importance to keep the data consistency, which is widely applied in many fields, such as banking, insurance, retail industry, telecommunications, and traffic control.

In e-commerce, most data is electronically collected, which may have serious data quality problems. Classic data quality problems mainly come from software defects, customized errors, or system mis-configuration. Authors in [42] discussed data cleaning in e-commerce by crawlers and regularly re-copying customer and account information. In [43], the problem of cleaning RFID data was examined. RFID is widely used in many applications, e.g., inventory management and target tracking. However, the original RFID features low quality, which includes a lot of abnormal data limited by the physical design and affected by environmental noises. In [44], a probability model was developed to cope with data loss in mobile environments. Khoussainova et al. in [45] proposed a system to automatically correct errors of input data by defining global integrity constraints. Herbert et al. [46]

proposed a framework called BIO-AJAX to standardize biological data so as to conduct further computation and improve search quality. With BIO-AJAX, some errors and repetitions may be eliminated, and common data mining technologies can be executed more effectively.

3.2.3.3 Redundancy Elimination

Data redundancy refers to data repetitions or surplus, which usually occurs in many datasets. Data redundancy can increase the unnecessary data transmission expense and cause defects on storage systems, e.g., waste of storage space, leading to data inconsistency, reduction of data reliability, and data damage. Therefore, various redundancy reduction methods have been proposed, such as redundancy detection, data filtering, and data compression. Such methods may apply to different datasets or application environments. However, redundancy reduction may also bring about certain negative effects. For example, data compression and decompression cause additional computational burden. Therefore, the benefits of redundancy reduction and the cost should be carefully balanced.

Data collected from different fields will increasingly appear in image or video formats. It is well-known that images and videos contain considerable redundancy, including temporal redundancy, spacial redundancy, statistical redundancy, and sensing redundancy. Video compression is widely used to reduce redundancy in video data, as specified in the many video coding standards (MPEG-2, MPEG-4, H.263, and H.264/AVC). In [47], the authors investigated the problem of video compression in a video surveillance system with a video sensor network. The authors propose a new MPEG-4 based method by investigating the contextual redundancy related to background and foreground in a scene. The low complexity and the low compression ratio of the proposed approach were demonstrated by the evaluation results.

On generalized data transmission or storage, repeated data deletion is a special data compression technology, which aims to eliminate repeated data copies [48]. With repeated data deletion, individual data blocks or data segments will be assigned with identifiers (e.g., using a hash algorithm) and stored, with the identifiers added to the identification list. As the analysis of repeated data deletion continues, if a new data block has an identifier that is identical to that listed in the identification list, the new data block will be deemed as redundant and will be replaced by the corresponding stored data block. Repeated data deletion can greatly reduce storage requirement, which is particularly important to a big data storage system. Apart from the aforementioned data pre-processing methods, specific data objects shall go through some other operations such as feature extraction. Such operation plays an important role in multimedia search and DNA analysis [49–51]. Usually high-dimensional feature vectors (or high-dimensional feature points) are used to describe such data objects and the system stores the dimensional feature vectors for future retrieval. Data transfer is usually used to process distributed heterogeneous data sources, especially business datasets [52].

As a matter of fact, in consideration of various datasets, it is non-trivial, or impossible, to build a uniform data pre-processing procedure and technology that is applicable to all types of datasets. on the specific feature, problem, performance requirements, and other factors of the datasets should be considered, so as to select a proper data pre-processing strategy.

References

1. James Manyika, McKinsey Global Institute, Michael Chui, Brad Brown, Jacques Bughin, Richard Dobbs, Charles Roxburgh, and Angela Hung Byers. *Big data: The next frontier for innovation, competition, and productivity*. McKinsey Global Institute, 2011.
2. John Gantz and David Reinsel. The digital universe decade-are you ready. *External Publication of IDC (Analyse the Future) Information and Data*, pages 1–16, 2010.
3. Douglas Laney. 3-d data management: Controlling data volume, velocity and variety. *META Group Research Note, February*, 6, 2001.
4. Kenneth Cukier. *Data, data everywhere: A special report on managing information*. Economist Newspaper, 2010.
5. Paul Zikopoulos, Chris Eaton, et al. *Understanding big data: Analytics for enterprise class hadoop and streaming data*. McGraw-Hill Osborne Media, 2011.
6. Randal E Bryant. Data-intensive scalable computing for scientific applications. *Computing in Science & Engineering*, 13(6):25–33, 2011.
7. Mohd Helmy Abd Wahab, Mohd Norzali Haji Mohd, Hafizul Fahri Hanafi, and Mohamad Farhan Mohamad Mohsin. Data pre-processing on web server logs for generalized association rules mining algorithm. *World Academy of Science, Engineering and Technology*, 48:2008, 2008.
8. Alexandros Nanopoulos, Yannis Manolopoulos, Maciej Zakrzewicz, and Tadeusz Morzy. Indexing web access-logs for pattern queries. In *Proceedings of the 4th international workshop on Web information and data management*, pages 63–68. ACM, 2002.
9. Karuna P Joshi, Anupam Joshi, and Yelena Yesha. On using a warehouse to analyze web logs. *Distributed and Parallel Databases*, 13(2):161–180, 2003.
10. Vijay Chandramohan and Ken Christensen. A first look at wired sensor networks for video surveillance systems. In *Local Computer Networks, 2002. Proceedings. LCN 2002. 27th Annual IEEE Conference on*, pages 728–729. IEEE, 2002.
11. Leo Selavo, Anthony Wood, Qing Cao, Tamim Sookoor, Hengchang Liu, Aravind Srinivasan, Yafeng Wu, Woochul Kang, John Stankovic, Don Young, et al. Luster: wireless sensor network for environmental research. In *Proceedings of the 5th international conference on Embedded networked sensor systems*, pages 103–116. ACM, 2007.
12. Guillermo Barrenetxea, François Ingelrest, Gunnar Schaefer, Martin Vetterli, Olivier Couach, and Marc Parlange. Sensorscope: Out-of-the-box environmental monitoring. In *Information Processing in Sensor Networks, 2008. IPSN'08. International Conference on*, pages 332–343. IEEE, 2008.
13. Younghun Kim, Thomas Schmid, Zainul M Charbiwala, Jonathan Friedman, and Mani B Srivastava. Nawms: nonintrusive autonomous water monitoring system. In *Proceedings of the 6th ACM conference on Embedded network sensor systems*, pages 309–322. ACM, 2008.
14. Sukun Kim, Shamim Pakzad, David Culler, James Demmel, Gregory Fenves, Steven Glaser, and Martin Turon. Health monitoring of civil infrastructures using wireless sensor networks. In *Information Processing in Sensor Networks, 2007. IPSN 2007. 6th International Symposium on*, pages 254–263. IEEE, 2007.
15. Matteo Ceriotti, Luca Mottola, Gian Pietro Picco, Amy L Murphy, Stefan Guna, Michele Corra, Matteo Pozzi, Daniele Zonta, and Paolo Zanon. Monitoring heritage buildings with wireless

sensor networks: The torre aquila deployment. In *Proceedings of the 2009 International Conference on Information Processing in Sensor Networks*, pages 277–288. IEEE Computer Society, 2009.

16. Gilman Tolle, Joseph Polastre, Robert Szewczyk, David Culler, Neil Turner, Kevin Tu, Stephen Burgess, Todd Dawson, Phil Buonadonna, David Gay, et al. A macroscope in the redwoods. In *Proceedings of the 3rd international conference on Embedded networked sensor systems*, pages 51–63. ACM, 2005.
17. Feng Wang and Jiangchuan Liu. Networked wireless sensor data collection: issues, challenges, and approaches. *Communications Surveys & Tutorials, IEEE*, 13(4):673–687, 2011.
18. Junghoo Cho and Hector Garcia-Molina. Parallel crawlers. In *Proceedings of the 11th international conference on World Wide Web*, pages 124–135. ACM, 2002.
19. Suryakant Choudhary, Mustafa Emre Dincturk, Seyed M Mirtaheri, Ali Moosavi, Gregor von Bochmann, Guy-Vincent Jourdan, and Iosif-Viorel Onut. Crawling rich internet applications: the state of the art. In *CASCON*, pages 146–160, 2012.
20. Nasir Ghani, Sudhir Dixit, and Ti-Shiang Wang. On ip-over-wdm integration. *Communications Magazine, IEEE*, 38(3):72–84, 2000.
21. James Manchester, Jon Anderson, Bharat Doshi, and Subra Dravida. Ip over sonet. *Communications Magazine, IEEE*, 36(5):136–142, 1998.
22. M Jinno, H Takara, and B Kozicki. Dynamic optical mesh networks: Drivers, challenges and solutions for the future. In *Optical Communication, 2009. ECOC'09. 35th European Conference on*, pages 1–4. IEEE, 2009.
23. Luiz André Barroso and Urs Hölzle. The datacenter as a computer: An introduction to the design of warehouse-scale machines. *Synthesis Lectures on Computer Architecture*, 4(1):1–108, 2009.
24. Jean Armstrong. Ofdm for optical communications. *Journal of lightwave technology*, 27(3): 189–204, 2009.
25. William Shieh. Ofdm for flexible high-speed optical networks. *Journal of Lightwave Technology*, 29(10):1560–1577, 2011.
26. Cisco data center interconnect design and deployment guide, 2010.
27. Albert Greenberg, James R Hamilton, Navendu Jain, Srikanth Kandula, Changhoon Kim, Parantap Lahiri, David A Maltz, Parveen Patel, and Sudipta Sengupta. Vl2: a scalable and flexible data center network. In *ACM SIGCOMM Computer Communication Review*, volume 39, pages 51–62. ACM, 2009.
28. Chuanxiong Guo, Guohan Lu, Dan Li, Haitao Wu, Xuan Zhang, Yunfeng Shi, Chen Tian, Yongguang Zhang, and Songwu Lu. Bcube: a high performance, server-centric network architecture for modular data centers. *ACM SIGCOMM Computer Communication Review*, 39(4):63–74, 2009.
29. Nathan Farrington, George Porter, Sivasankar Radhakrishnan, Hamid Hajabdolali Bazzaz, Vikram Subramanya, Yeshaiahu Fainman, George Papen, and Amin Vahdat. Helios: a hybrid electrical/optical switch architecture for modular data centers. *ACM SIGCOMM Computer Communication Review*, 41(4):339–350, 2011.
30. Hussam Abu-Libdeh, Paolo Costa, Antony Rowstron, Greg O'Shea, and Austin Donnelly. Symbiotic routing in future data centers. *ACM SIGCOMM Computer Communication Review*, 40(4):51–62, 2010.
31. Cedric Lam, Hong Liu, Bikash Koley, Xiaoxue Zhao, Valey Kamalov, and Vijay Gill. Fiber optic communication technologies: What's needed for datacenter network operations. *Communications Magazine, IEEE*, 48(7):32–39, 2010.
32. Guohui Wang, David G Andersen, Michael Kaminsky, Konstantina Papagiannaki, TS Ng, Michael Kozuch, and Michael Ryan. c-through: Part-time optics in data centers. In *ACM SIGCOMM Computer Communication Review*, volume 40, pages 327–338. ACM, 2010.
33. Xiaohui Ye, Yawei Yin, SJ Ben Yoo, Paul Mejia, Roberto Proietti, and Venkatesh Akella. Dos: A scalable optical switch for datacenters. In *Proceedings of the 6th ACM/IEEE Symposium on Architectures for Networking and Communications Systems*, page 24. ACM, 2010.

34. Ankit Singla, Atul Singh, Kishore Ramachandran, Lei Xu, and Yueping Zhang. Proteus: a topology malleable data center network. In *Proceedings of the 9th ACM SIGCOMM Workshop on Hot Topics in Networks*, page 8. ACM, 2010.
35. Odile Liboiron-Ladouceur, Isabella Cerutti, Pier Giorgio Raponi, Nicola Andriolli, and Piero Castoldi. Energy-efficient design of a scalable optical multiplane interconnection architecture. *Selected Topics in Quantum Electronics, IEEE Journal of*, 17(2):377–383, 2011.
36. Avinash Karanth Kodi and Ahmed Louri. Energy-efficient and bandwidth-reconfigurable photonic networks for high-performance computing (hpc) systems. *Selected Topics in Quantum Electronics, IEEE Journal of*, 17(2):384–395, 2011.
37. Xia Zhou, Zengbin Zhang, Yibo Zhu, Yubo Li, Saipriya Kumar, Amin Vahdat, Ben Y Zhao, and Haitao Zheng. Mirror mirror on the ceiling: Flexible wireless links for data centers. *ACM SIGCOMM Computer Communication Review*, 42(4):443–454, 2012.
38. Maurizio Lenzerini. Data integration: A theoretical perspective. In *Proceedings of the twenty-first ACM SIGMOD-SIGACT-SIGART symposium on Principles of database systems*, pages 233–246. ACM, 2002.
39. Wiki. Applications and organizations using hadoop. http://wiki.apache.org/hadoop/PoweredBy, 2013.
40. Michael J Cafarella, Alon Halevy, and Nodira Khoussainova. Data integration for the relational web. *Proceedings of the VLDB Endowment*, 2(1):1090–1101, 2009.
41. Jonathan I Maletic and Andrian Marcus. Data cleansing: Beyond integrity analysis. In *IQ*, pages 200–209. Citeseer, 2000.
42. Ron Kohavi, Llew Mason, Rajesh Parekh, and Zijian Zheng. Lessons and challenges from mining retail e-commerce data. *Machine Learning*, 57(1–2):83–113, 2004.
43. Haiquan Chen, Wei-Shinn Ku, Haixun Wang, and Min-Te Sun. Leveraging spatio-temporal redundancy for rfid data cleansing. In *Proceedings of the 2010 ACM SIGMOD International Conference on Management of data*, pages 51–62. ACM, 2010.
44. Zhou Zhao and Wilfred Ng. A model-based approach for rfid data stream cleansing. In *Proceedings of the 21st ACM international conference on Information and knowledge management*, pages 862–871. ACM, 2012.
45. Nodira Khoussainova, Magdalena Balazinska, and Dan Suciu. Probabilistic event extraction from rfid data. In *Data Engineering, 2008. ICDE 2008. IEEE 24th International Conference on*, pages 1480–1482. IEEE, 2008.
46. Katherine G Herbert and Jason TL Wang. Biological data cleaning: a case study. *International Journal of Information Quality*, 1(1):60–82, 2007.
47. Tsung-Han Tsai and Chung-Yuan Lin. Exploring contextual redundancy in improving object-based video coding for video sensor networks surveillance. *Multimedia, IEEE Transactions on*, 14(3):669–682, 2012.
48. Sunita Sarawagi and Anuradha Bhamidipaty. Interactive deduplication using active learning. In *Proceedings of the eighth ACM SIGKDD international conference on Knowledge discovery and data mining*, pages 269–278. ACM, 2002.
49. Uday Kamath, Jack Compton, Rezarta Islamaj Dogan, Kenneth De Jong, and Amarda Shehu. An evolutionary algorithm approach for feature generation from sequence data and its application to dna splice site prediction. *IEEE/ACM Transactions on Computational Biology and Bioinformatics (TCBB)*, 9(5):1387–1398, 2012.
50. Kwong-Sak Leung, Kin Hong Lee, Jin-Feng Wang, Eddie YT Ng, Henry LY Chan, Stephen KW Tsui, Tony SK Mok, PC-H Tse, and JJ-Y Sung. Data mining on dna sequences of hepatitis b virus. *Computational Biology and Bioinformatics, IEEE/ACM Transactions on*, 8(2):428–440, 2011.
51. Zi Huang, Hengtao Shen, Jiajun Liu, and Xiaofang Zhou. Effective data co-reduction for multimedia similarity search. In *Proceedings of the 2011 ACM SIGMOD International Conference on Management of data*, pages 1021–1032. ACM, 2011.
52. Jens Bleiholder and Felix Naumann. Data fusion. *ACM Computing Surveys (CSUR)*, 41(1):1, 2008.

Chapter 4
Big Data Storage

Abstract In this chapter, we focus on the storage of big data. We will review important issues including massive storage systems, distributed storage systems, and big data storage mechanisms. On one hand, the storage infrastructure need to provide information storage service with reliable storage space; on the other hand, it must provide a powerful access interface for query and analysis of large amount of data. Such a storage infrastructure generally consists of hardware infrastructure and storage mechanisms.

4.1 Storage System for Massive Data

Data storage refers to the storage and management of large-scale datasets, while achieving reliability and availability. A data storage system consists of two parts: infrastructure and data storage methods or mechanisms. The hardware infrastructure includes massive shared Information Communication Technology (ICT) resources utilized to feedback instant demands of tasks, and such ICT resources are organized in an elastic manner. The hardware infrastructure shall feature elasticity and dynamic reconfiguration, to be adaptive to different application environments. Data storage methods are deployed on the top of hardware infrastructure to maintain large-scale datasets. Storage systems shall be equipped with many interfaces, rapid query, or other programming models for analysis of or interaction with stored data.

The big data paradigm brings about an explosive growth of data. The sharp growth of data has stringent requirements on storage and management. Traditionally, data storage equipment is only auxiliary equipment of servers, and most of them store, manage, look up, and analyze data with structured RDBMSs. The GB, TB, to PB sharp growth of big data makes traditional storage equipment and management modes inadequate. Data storage equipment is becoming increasingly more important, and storage cost becomes the main expense of many Internet companies. Therefore, there is a compelling need for research on data storage.

M. Chen et al., *Big Data: Related Technologies, Challenges and Future Prospects*, SpringerBriefs in Computer Science, DOI 10.1007/978-3-319-06245-7_4,

A large number of storage systems emerge to meet the demands of big data. Existing storage technologies can be classified as DAS (Direct Attached Storage) and network storage, while network storage can be further classified into NAS (Network Attached Storage) and SAN (Storage Area Network).

In DAS, disc drives are directly connected with servers. Storage is a peripheral equipment, while, data management servers and all kinds of application software are matched with storage sub-systems (this way, I/O may stress system bandwidths). DAS applies to a few server environments but, when the storage capacity is increased, the efficiency of storage supply will be quite low and the upgradeability and expandability will be greatly limited. In case of server abnormality, data could not be acquired and stored resources and data could not be shared. DAS is mainly used in personal computers and small-sized servers, which only support such applications requiring low storage capacities and does not directly support multi-computer shared storage. Tap drivers and RAID (redundant array of independent disks) are classic DAS equipments.

Network storage is to utilize the original network or a specially designed storage network to provide users with a uniform information access and sharing services of information systems. Network storage equipment includes special data exchange equipments, disk array, tap library, and other storage media, as well as special storage software. It is characterized with mass data storage, limited data sharing, full utilization of data mining and information, data reliability, data backup and safety, as well as simplified and unified data management. In addition, network storage features very strong expandability, so as to provide information transmission rates suited for large data volume.

NAS is actually an auxiliary storage equipment of a network. It is directly connected to a network through a hub or switch, communicating with the TCP/IP protocol. NAS is geared to message passing, and transmits data in the form of files. NAS has two prominent features. First, on physical connection, it directly connects the storage equipment to a network and then hangs the storage at the rear end of a server, thus avoiding the I/O burden at the server. Second, technically, it reduces the movements of the disk actuator arm and thus reduces R/W delay. However, the structure of NAS shows that it is still a traditional server storage equipment in essence.

SAN focuses on data storage with a flexible network topology and high-speed optical fiber connections. It allows multipath data switching among any internal nodes. Data storage management is located in a relatively independent storage local area network, so as to achieve a maximum degree of data sharing and data management, as well as seamless extension of the system. From the organization of a data storage system, DAS, NAS, and SAN can all be divided into three parts: (a) disk array: it is the foundation of a storage system and the fundamental guarantee for data storage; (b) connection and network sub-systems, which provide connection among one or more disc arrays and servers; (c) storage management software, which handles data sharing, disaster recovery, and other storage management tasks of multiple servers.

4.2 Distributed Storage System

The first challenge brought about by big data is how to develop a large scale distributed storage system for strategic preservation of data and efficient data processing and analysis. To use a distributed system to store massive data, the following factors should be taken into consideration:

- *Consistency*: a distributed storage system requires multiple servers to cooperatively store data. As there are more servers, the probability of server failures will be larger. Usually data is divided into multiple pieces to be stored at different servers to ensure availability in case of server failure. However, server failures and parallel storage may cause inconsistency among different copies of the same data. Consistency refers to assuring that multiple copies of the same data are identical.
- *Availability*: a distributed storage system operates in multiple sets of servers. As more servers are used, server failures are inevitable. It would be desirable if the entire system is not serious affected with respect to serving the reading and writing requests from customer terminals. This property is called availability.
- *Partition Tolerance*: multiple servers in a distributed storage system are connected by a network. The network could have link/node failures or temporary congestion. The distributed system should have a certain level of tolerance to problems caused by network failures. It would be desirable that the distributed storage still works well when the network is partitioned.

Eric Brewer proposed a CAP [1, 2] theory in 2000, which indicated that a distributed system could not simultaneously meet the requirements on consistency, availability, and partition tolerance; at most two of the three requirements can be satisfied simultaneously. Seth Gilbert and Nancy Lynch from MIT proved the correctness of CAP theory in 2002. Since consistency, availability, and partition tolerance could not be achieved simultaneously, we can have a CA system by ignoring partition tolerance, a CP system by ignoring availability, and an AP system that ignores consistency, according to different design goals. The three systems are discussed in the following.

CA systems do not have partition tolerance, i.e, they could not handle network failures. Therefore, CA systems are generally deemed as storage systems with a single server, such as the traditional small-scale relational databases. Such systems feature single copy of data, such that consistency is easily ensured. Availability is guaranteed by the excellent design of relational databases. However, since CA systems could not handle network failures, they could not be expanded to use many servers. This is way most large-scale storage systems are CP systems and AP systems.

Compared with CA systems, CP systems ensure partition tolerance. Therefore, CP systems can be expanded to become distributed systems. CP systems generally maintain several copies of the same data in order to ensure a level of fault tolerance. CP systems also ensure data consistency, i.e., multiple copies of the same data are guaranteed to be completely identical. However, CP could not ensure sound availability because of the high cost for consistency assurance. Therefore, CP systems are useful for the scenarios with moderate load but stringent requirements on data accuracy (e.g., trading data). BigTable and Hbase are two popular CP systems. BigTable is well-known since it was successful for managing the background data of Google's search engine. Because a lot of data in Google is structured data, BigTable mainly stores data with tables. Nevertheless, when a lot of information is put in a table, the table size will grow. Such information should be partitioned and stored separately. The table is usually highly sparse. Therefore, BigTable divides the columns into different Column Families, where every column family stores the same type of information. This way, similar data is stored together and the same type of information is processed in the same manner, making it easy for system users. In the same column family, new columns can be arbitrarily inserted, thus reducing the usage limit of BigTable to a great extent.

BigTable is designed in the way similar to GFS, a distributed file system of Google, where one Master and several Tablet Servers constitute a star structure in a system. The star structure has a single point of failure. The load of the Master server should be reduced in order to minimize Master errors. In BigTable, data transmission and data addressing do not involve the Master. Therefore the load of the Master is not high. In order to solve the problem of a single point of failure, BigTable adopts a Master election mechanism. In particular, it incorporates an asynchronous and consistent locking mechanism to ensure that exact one Master is elected every time based on the Paxos protocol [3].

Data in BigTable is sequenced in the lexicographic order of rows. During data modification, we shall insert a record in a sequential table, find a position to be inserted, and then move the original data to make room for the newly inserted data. This operation is very time-consuming. BigTable utilizes batch processing to solve this problem. Specifically, BigTable uses two tables to store data: it stores historical data with a big table and stores recently modified data with a very small table.when the recent data accumulates to a certain amount or after a certain amount of time, BigTable merges the recent data into the historical data. This approach greatly reduces the times that big tables are modified, since only small tables are frequently modified. The cost of data modification is thus reduced to a great extent. Therefore, this method mitigates the problem of high cost for data changes and increases the look-up speed for recently modified data.

AP systems,also ensure partition tolerance. However, AP systems are different from CP systems in that AP systems also ensure availability. However, AP systems only ensure eventual consistency rather than strong consistency in the previous two systems. Therefore, AP systems only apply to the scenarios with frequent requests but not very high requirements on accuracy. For example, in online SNS (Social Networking Services) systems, there are many concurrent visits to the data but certain amount of data errors are tolerable. Furthermore, because AP systems ensure eventual consistency, accurate data can still be obtained after a certain amount of delay. Therefore, AP systems may also be used under the circumstances with no stringent real-time requirements.

Dynamo and Cassandra are two popular AP systems. Cassandra, with sound expandability, is used for storing massive textual data by mainstream commercial online SNS companies, such as Facebook and Twitter. Specifically, Cassandra utilizes the Consistent Hash algorithm to randomly and evenly map Key spaces of user identifier spaces of servers to the same value domain space, and enables servers to manage user data corresponding to the Keys of adjacent mapped values. This way, dynamic changes at any servers in the systems only affect the data corresponding to a small segment of value domains undertaken by themselves. Mainstream SNSs utilize such distributed Key-Value storage approach, so as to better meet the demands of expandability of large-scale online SNS systems and load balance for the servers, and be adaptive to dynamic changes of systems.

In order to support the storage of textual data of users, Cassandra inherits the column family model of BigTable to aggregate data with similar features into a column family. What is different from BigTable is that Cassandra may expand the concept of column family to a super column family-the column family of column families. On Cassandra nodes, every column family corresponds to a MemTable, a resident memory. When nodes write data, it first writes data into MemTable. In proper occasions, e.g., memory space occupied by MemTable reaches the upper bound or after a fixed amount of time, MemTable is stored into a corresponding SSTable of a disk. SSTable has a large operation throughput because of its sequential writing approach. The system builds a local index for every block including every piece of data written in the disk and then Cassandra stores the index in the internal memory in the form of Bloom Filter compression [4]. Because the compressed index excludes the relative position of the block in the file system, Cassandra does not perform well with regard to random reading.

4.3 Storage Mechanism for Big Data

Considerable research on big data promotes the development of storage mechanisms for big data. Existing storage mechanisms of big data may be classified into three bottom-up levels: (a) file systems, (b) databases, and (c) programming models.

File systems are the foundation of the applications at upper levels. Google's GFS is an expandable distributed file system to support large-scale, distributed, data-intensive applications [5]. GFS uses cheap commodity servers to achieve fault-tolerance and provides customers with high-performance services. GFS supports large-scale file applications with more frequent reading than writing. However, GFS also has some limitations, such as a single point of failure and poor performances for small files. Such limitations have been overcome by Colossus [6], the successor of GFS.

In addition, other companies and researchers also have their solutions to meet the different demands for storage of big data. For example, HDFS and Kosmosfs are derivatives of open source codes of GFS. Microsoft developed Cosmos [7] to support its search and advertisement business. Facebook utilizes Haystack [8] to store the large amount of small-sized photos. Taobao also developed TFS and FastDFS. In conclusion, distributed file systems have been relatively mature after years of development and business operation. Therefore, we will focus on the other two levels in the rest of this section.

4.3.1 Database Technology

The database technology has been evolving for more than 30 years. Various database systems are developed to handle datasets at different scales and support various applications. It is apparent that traditional relational databases cannot meet the challenges on categories and scales brought about by big data. NoSQL databases (i.e., non traditional relational databases) are becoming more popular for big data storage. NoSQL databases feature flexible modes, support for simple and easy copy, simple API, eventual consistency, and support of large volume data. NoSQL databases are becoming the core technology for of big data. We will examine the following three main NoSQL databases in this section: Key-value databases, column-oriented databases, and document-oriented databases, each based on certain data models.

4.3.1.1 Key-Value Databases

Key-value Databases are constituted by a simple data model and data is stored corresponding to key-values. Every key is unique and customers may input queried values according to the keys. Such databases feature a simple structure and the modern key-value databases are characterized with high expandability and smaller query response time higher than those of relational databases. Over the past few years, many key-value databases have appeared as motivated by Amazon's Dynamo system [9]. We will next introduce Dynamo and several other representative key-value databases.

Dynamo

Dynamo is a highly available and expandable distributed key-value data storage system. It is used to manage store status of some core services in the Amazon e-Commerce Platform. Amazon e-Commerce Platform provides multiple services and data storage that can be realized with key access. The public mode of relational databases may generate invalid data and limit data scale and availability. Dynamo can meet requirements of such applications with a simple key-object interface. The Dynamo interface is constituted by simple reading and writing of data items. Dynamo achieves elasticity and availability through the data partition, data copy, and object edition mechanisms.

The Dynamo partition plan relies on Consistent Hashing [10] to divide load for multiple main storage machines. With this mechanism, the mapping scope of a hash function is deemed as a fixed circular space or "ring" (i.e. the maximum hash value is followed by the minimum hash value). Every node in the system is assigned with a random value in the space and such random value represents its "position" in the ring. The position of every data item identified with a key, can be computed through the calculation of hash value of a keyword in the data item. Then, we get the first node clockwise, with the position larger than that of the data item. This way, every node shall be responsible for the region between the node and the previous node. The Consistent Hash has a main advantage that node passing only affects directly adjacent nodes and do not affect other nodes.

Dynamo copies data items to N sets of mainframe computers, in which N is a configurable parameter in order to achieve high availability and durability. It distributes every key word K into a coordinating node. The coordinating node is responsible for the copy of data items within its scope. Apart from storing all key words within its scope, the coordinating node shall copy N−1 successive nodes in the ring clockwise. This way, every node in the system will be responsible for a region between itself and the Nth former node.

Dynamo system also provides eventual consistency, so as to conduct asynchronous update on all copies. Before the update is applied to all copies, the *put()* call may return to the caller. Consequently, the data returned by the next *get()* call may not be the recently updated data. If there is no failure, the propagation delay of updating can be determined. However, in case of failure (e.g, server failures or network partition), the update may not propagate to all the copies until a large delay.

Voldemort

Voldemort is also a key-value storage system, which was initially developed for and is still used by LinkedIn. Key words and values in Voldemort are composite objects constituted by tables and images. The voldemort interface includes three simple operations: reading, writing, and deletion, all of which are confirmed by key words. Voldemort provides asynchronous updating concurrent control of multiple editions but does not ensure data consistency. However, Voldemort supports optimistic locking for consistent multi-record updating: in case of conflict between the updating and any other operations, the updating operation may exit. The vector clock used in Dynamo provides various data editions with sequencing. The data copy mechanism of Voldmort is the same as that of Dynamo. Voldemort may store data in RAM but allows data be inserted into a storage engine. It is worth noting that Voldemort supports two storage engines including Berkeley DB and Random Access Files.

The key-value database emerged a few years ago. Deeply influenced by Amazon DynamoDB, other key-value storage systems include Redis, Tokyo Canbinet and Tokyo Tyrant, Memcached and MemcacheDB, Riak and Scalaris, all of which provide expandability by distributing key words into nodes. Voldemort, Riak, Tokyo

Cabinet, and Memecached can utilize attached storage devices to store data in RAM or disks. Other storage systems store data at RAM and provide disc backup, or rely on copies and copy recovery to avoid the need for backup.

4.3.1.2 Column-Oriented Databases

The column-oriented databases store and process data according to columns other than rows. Columns and rows are segmented in multiple nodes to realize expandability. The column-oriented databases are mainly inspired by Google's BigTable. In this section, we first discuss BigTable and then introduce several derivative tools.

BigTable

BigTable is a distributed, structured data storage system, which is designed to process the large-scale (PB class) data among thousands commercial servers [3]. The basic data structure of BigTable is a multi-dimension sequenced mapping with sparse, distributed, and persistent storage. Indexes of mapping are key words of rows, key words of columns, and timestamps, and every value in mapping is an unanalyzed byte array. The key words of rows in BigTable are 64KB character strings, in which the rows are stored according to the lexicographical order and are continually segmented into Tablets, i.e. units of distribution and load balance. This way, read a short row of data can be highly effective, since it only involves communication with a small portion of machines. The columns are grouped according to the prefixes of key words, which are called column families. These column families are the basic units for access control. The timestamps are 64-bit integers to distinguish different editions of cell values. Clients may flexibly determine the quantity of cell editions to be stored. These editions are sequenced in the descending order of timestamps, so the latest edition will always be read.

The BigTable API features the creation and deletion of Tablets and column families as well as modification of metadata of clusters, tables, and column families, and access control rights. Client applications may write or delete values of BigTable, look up values from columns, or browse sub-datasets in a table. BigTable also supports some other characteristics, such as transaction processing in a single row. Users may utilize such features to conduct more complex processing on data.

BigTable is based on many fundamental components of Google, including GFS [5], cluster management system, SSTable file format, and Chubby [11]. GFS is use to store data and log files. The cluster management system is responsible for task scheduling, management of shared resources in machines, processing of machine failures, and monitoring of machine statuses. SSTable file format is used to store BigTable data internally. SStable provides mapping between persistent, sequenced, and unchangeable key words and values, with key words and values of any byte strings. BigTable utilizes Chubby for the following server tasks: (1) ensure there is at most one active Master copy at any time; (2) store the bootstrap location of

BigTable data; (3) look up Tablet server; (4) conduct error recovery in case of Table server failures; (5) store BigTable schema information; (6) store the access control table.

Every procedure executed by BigTable includes three main components: Master server, Tablet server, and client library. BigTable only allows one set of Master server be distributed to be responsible for distributing tablets for Tablet server, detecting added or removed Tablet servers, and conducting load balance. In addition, it can also modify the BigTable schema, e.g., creating tables and column families, and collecting garbage saved in GFS as well as deleted or disabled files, and using them in specific BigTable instances. Every tablet server manages a Tablet set and is responsible for the processing of loaded Tablet reading and writing, and segmenting Tablets when they are too big. The companying application client library is used to communicate with BigTable instances.

Cassandra

Cassandra is a distributed storage system to manage the huge amount of structured data distributed among multiple commercial servers [12]. The system was developed by Facebook and became an open source tool in 2008. It adopts the ideas and concepts of both Amazon Dynamo and Google BigTable, especially integrating the distributed system technology of Dynamo with the BigTable data model. Tables in Cassandra are in the form of distributed four-dimensional structured mapping, where the four dimensions including row, column family, column, and super column. A row is distinguished by a string-key with arbitrary length. No matter what the amount of columns to be read or written is, the operation on rows is an atomic operation. Columns may constitute clusters, which is called column families, which are similar to the data model of BigTable. Cassandra provides two kinds of column families: column families and super columns. The super column includes any quantity of columns with names related to the super column. A column family includes columns and super columns, which may be continuously added to the column family during execution. The partition and copy mechanisms of Cassandra are very similar to those of Dynamo, so as to achieve consistency.

Derivative Tools of BigTable

Since the BigTable code cannot be obtained through the open source license, some open source projects compete to implement the BigTable concept to develop similar systems, such as HBase and Hypertable.

HBase is a BigTable clone programmed with Java and is a part of Hadoop of Apache's MapReduce framework [13]. HBase replaces GFS with HDFS. It writes updated contents into the RAM and regularly writes them into files in discs. The row operations are atomic operations, equipped with row-level locking and transaction processing. Large-scope transaction processing is provided with optional support. Partition and distribution are transparently operated, with client hash or a fixed secret key space.

HyperTable was developed similar to BigTable to obtain a set of high-performance, expandable, and distributed storage and processing systems for structured and unstructured data [14]. HyperTable relies on distributed file systems, e.g. HDFS, and distributed lock manager. Data representation, processing, and partition mechanism are similar to that in BigTable. HyperTable has its own query language, called HyperTable query language (HQL), and allows users to create, modify, and query underlying tables.

Since the column-oriented storage databases mainly emulate BigTable, their designs are all similar, except for the concurrency mechanism and several other features. For example, Cassandra emphasizes weak concurrency of concurrent control of multiple editions, while HBase and HyperTable focus on strong consistency through locks or log records.

4.3.1.3 Document Databases

Compared with key-value storage, document storage can support more complex data forms. Since documents do not follow strict modes, there is no need to conduct mode migration. In addition, key-value pairs can still be saved. We will examine three important representatives of document storage systems, i.e., MongoDB, SimpleDB, and CouchDB.

MongoDB

MongoDB is an open-source document-oriented database [15]. MongoDB stores documents as Binary JSON (BSON) objects [16], which is similar to object. Every document has an ID field as the main key word. Query in MongoDB is expressed with syntax similar to JSON. A database driver sends the query as a BSON object to MongoDB. The system allows query on all documents, including embedded objects and arrays. Indexes may be created for queryable fields in documents to enable rapid query.

The copy operation in MongoDB can be executed with log files in the main nodes that support all the high-level operations conducted in the database. During the copy operation, the machine queries all the writing operations since the last synchronization of the machine and executing operations in log files in local databases. MongoDB supports horizontal expansion with automatic sharing to distribute data among thousands of nodes by automatically balancing load and keep the system up and running in case of failure.

SimpleDB

SimpleDB is a distributed database and a web service of Amazon [17]. Data in SimpleDB is organized into various domains in which data may be stored, acquired, and queried. Domains include different properties and name/value pair sets of projects. Date is copied to different machines at different data centers in order to ensure data safety and improve performance. This system does not support automatic partition and thus could not be expanded with the change of data volume. SimpleDB allows users to use SQL to run query, e.g., selecting sentences nonconforming to a single domain. It is worth noting that SimpleDB can assure eventual consistency but does not feature MVCC (Muti-Version Concurrency Control). Therefore, conflicts therein could not be detected from the client side.

CouchDB

Apache CouchDB is a document-oriented database written in Erlang [18]. Data in CouchDB is organized into documents that consist of fields named by key words/names and values, and are stored and accessed as JSON objects. Every document is provided with a unique identifier. CouchDB allows access to database documents through the RESTful HTTP API. If a document needs to be modified, the client can download the entire document, modify it, and then send it back to the database. After a document is rewritten once, the identifier will be modified and updated. CouchDB utilizes the optimal copying to acquire scalability but without a sharing mechanism. Since various CouchDBs may be executed along with other transactions simultaneously, any kinds of Replication Topology can be built. The consistency type of CouchDB relies on the copy mechanism. If the server-server configuration is utilized, CouchDB system can ensure eventual consistency; with the master-slave configure, strong consistency can be assured. MVCC in CouchDB is synchronously executed with the historical Hash records.

Except for properties, sets, and indexes defined in sets, all documents are stored without a document schema. Generally, they do not provide explicit lock and, compared with traditional relational databases, feature weaker concurrency and atomic properties. Documents may be distributed to nodes of all systems to achieve scalability at different levels.

4.3.1.4 Platform for Nimble Universal Table Storage

Platform for Nimble Universal Table Storage (PNUTS) is a large-scale parallel geographically-distributed system for Yahoo!'s web applications [19]. It relies on a simple relational data model in which data is organized into a property record table. In addition to the classic data types, blob (binary large object or basic large object) is also an effective data type that allows any structures within records (not always large-scale binary objects such as images or audio frequency).

In the physical layer, the system is divided into different regions, each of which includes a set of complete system components and complete copies of tables. The data table is horizontally segmented into record groups, which are called Tablets. Tablets are distributed among many servers; every server may have tens of thousands of Tablets but a Tablet may only be stored in a region of a server. The query language of PNUTS supports the selection and projection of a signal table. To upgrade or delete an existing record, the main key words must be specified. The consistency mode provided by PNUTS has a feature between the general serializability and eventual consistency.

4.3.2 Design Factors

Of the various database systems, there is not a single system that can achieve the optimal performance under all workload circumstances. In each database system, some performance goals have to be compromised to achieve optimized operation for specific applications.

Cooper et al. in [20] discussed the trade-offs confronted by data management systems based on cloud computing, including reading performance and writing performance, delay and durability, synchronous and asynchronous copies, and data segmentation, among others. Some researchers also differentiated and analyzed other design factors [21–23]. In the following, we compare several prominent features of the existing database systems (rather than analyzing the design goals of a specific system).

- *Data Model*: this section examined three core data models, i.e. key-value, column, and document models. In particular, PNUTS uses a row-oriented data model.
- *Data Storage*: in some systems data are designed to be stored in RAM and their snapshots or copies are stored in discs. Other systems store data in discs, with the cache stored in RAM. A few systems have pluggable background programs that are allowed to use different data storage media, or standardized underlying document systems are required.
- *Concurrency Control*: there are three concurrency control mechanisms used in the existing systems: lock, MVCC, and non-concurrency control. The lock mechanism only allows a user to read or modify a real object (i.e., object, document, or row) at any time. The MVCC mechanism ensures the reading consistency. However, if several users modify a real object at the same time, several conflicting editions of a real object may be created. Some systems do not offer atomicity but allow different users to concurrently modify different parts of the same object, and may not ensure which edition will be acquired during reading.
- *Consistency*: according to the CAP theorem, strict consistency could not be simultaneously achieved along with availability and partition tolerance. The

weak consistency, eventual consistency, and time axis consistency of both types should be generally compromised for each other. Eventual consistency means that all the updating operations will finally be propagated through the system and all the copies will be eventually consistent beyond a given period of time. Time axis consistency means that all the copies of a given record will apply the updating operations following the same order.

- *CAP Option*: The CAP theorem indicates that a shared data system may achieve at most two properties, among consistency, availability, and partition tolerance. Databases based on cloud computing needs to copy data from different servers in order to handle system failure in some regions, which basically requires consistency and availability. This way, the trade-off between consistency and availability can be determined. At present, various weak consistency models [24] have been proposed to achieve reasonable system availability.

4.3.3 Database Programming Model

The massive datasets of big data are generally stored in hundreds and even thousands of commercial servers. Apparently, the traditional parallel models (e.g., Message Passing Interface (MPI) and Open Multi-Processing (OpenMP)) may not be adequate to support such large-scale parallel programs.

Some parallel programming modes have been proposed for specific fields. These models effectively improve the performance of NoSQL and reduce the performance gap between relational databases. Therefore, these models have become the cornerstone for the analysis of massive data.

4.3.3.1 MapReduce

MapReduce [25] is a simple but powerful programming model for large-scale computing using a large number of clusters of commercial PCs to achieve automatic parallel processing and distribution. In MapReduce, the computational workload are caused by inputting key-value pair sets and generating key-value pair sets. The computing model only has two functions, i.e., Map and Reduce, both of which are programmed by users. The Map function processes input and generates intermediate key-value pairs. Then, MapReduce will combine all the intermediate values related to the same key and transmit them to the Reduce function. Next, the Reduce function receives the intermediate key and its value set, merges them, and generates a smaller value set. MapReduce has the advantage that it avoids the complicated steps for developing parallel applications, e.g., data scheduling, fault-tolerance, and inter-node communications. The user only needs to program the two functions to develop a parallel application. The initial MapReduce framework did not support multiple datasets in a task. This shortcoming has been mitigated by some recent enhancements [26, 27].

Over the past decades, people have widely utilized the traditional relational databases to manage datasets. Consequently, programmers are familiar with the advanced declarative language of SQL, a relational database, for task description and dataset analysis. However, the succinct MapReduce framework only provides two nontransparent functions without the common operations (e.g., projects and filters). Therefore, programmers have to spend time on programming the basic functions, which are generally hard to maintain and reuse. Incorporating the SQL style in the MapReduce framework would be a promising solution. To this end, some advanced language systems have been proposed, e.g., the Sawzall [28] of Google, the Pig Latin [29] of Yahoo!, the Hive [30] of Facebook, and the Scope [3] of Microsoft, so as to improve the programming efficiency and user friendliness.

4.3.3.2 Dryad

Dryad [31] is a general-purpose distributed execution engine for processing parallel applications of coarse-grained data. The operational structure of Dryad is a directed acyclic graph, in which vertexes represent programs and edges represent data channels. Dryad executes operations on the vertexes in computer clusters and transmits data via data channels, including documents, TCP connections, and shared-memory FIFO. During operation, resources in a logic operation graph are automatically map to physical resources.

The operation structure of Dryad is coordinated by a central program called job manager, which can be executed in clusters or workstations of users. The user workstations can access clusters through the network. A job manager includes application codes and program library codes, in which application codes are used to build a job communication graph and the program library codes are used to arrange available resources. All kinds of data are directly transmitted between vertexes. Therefore, the job manager is only responsible for decision-making, which does not obstruct any data transmission.

In Dryad, application developers can flexibly choose any directed acyclic graph to describe the communication modes of the application and express data transmission mechanisms. In addition, Dryad allows vertexes to use any amount of input and output data, while MapReduce supports limited computing, with only one input set and generating only one output set. DryadLINQ [32] is the advanced language of Dryad and is used to integrate the aforementioned SQL-like language execution environment.

4.3.3.3 All-Pairs

All-Pairs [33] is a system specially designed for biometrics, bio-informatics, and data mining applications. It focuses on comparing element pairs in two datasets by a given function. The All-Pairs problem may be expressed as a three-tuples (Set A, Set B, and Function F), in which Function F is utilized to compare all elements in

Set A and Set B. The comparison result is an output matrix **M**. It is also called the Cartesian product or cross join of Set A and Set B.

All-Pairs is implemented in four phases: system modeling, input data distribution, batch job management, and result collection. In Phase I, an approximation model of system performance will be built to assist in deciding how much CPU is needed and how to conduct job partition. In Phase II, a spanning tree is built for data transmission, which is completed within logarithmic time. This way, the workload of every partition may effectively get input data. In Phase III, after the data flow is delivered to proper nodes, the All-Pairs engine will build a batch-processing submission for jobs in partitions, sequence them in the batch processing system, and formulate a node running command to acquire data in nodes. The user-defined items will be executed in proper partition jobs to generate results in batch, with the results displayed in the output matrix. In the last phase, as the batch processing system completes its jobs, the extraction engine will collect results and combine them in a proper structure, which is generally a single file list. In the list, all results are put in order. If the system hides the execution details from the user, the user may define the data and calculation requirements using the given interfaces.

4.3.3.4 Pregel

The Pregel [34] system of Google facilitates the processing of large-sized graphs, e.g., analysis of network graphs and social networking services. A computational task is expressed by a directed graph constituted by vertexes and directed edges, in which every vertex is related to a modifiable and user-defined value. Directed edges are related to their source vertexes and every edge is constituted by a modifiable and user-defined value and an identifier of a target vertex. After the graph is built, the program conducts iterative calculations, which is called supersteps among which global synchronization points are set until algorithm completion and output completion. In every superstep, vertex computations are parallel and every vertex executes the same user-defined function to express a given algorithm logic. Every vertex may modify its status and the status of its output edges, receive a message sent from the previous superstep, send the message to other vertexes, and even modify the topological structure of the entire graph. Edges are not provided with corresponding computations. Functions of every vertex may be removed by suspension. When all vertexes are in an inactive status without any message to transmit, the entire program execution is completed. The Pregel program output is a set consisting of the values output from all the vertexes. Generally speaking, the Pregel program output and input are an isomorphic directed graph.

Inspired by the aforementioned programming models, other researches have also focused on programming modes for more complex computational tasks, e.g., iterative computations [35, 36], fault-tolerant memory computations [37], incremental computations [38], and flow control decision-making related to data [39].

References

1. Eric A Brewer. Towards robust distributed systems. In *PODC*, page 7, 2000.
2. Seth Gilbert and Nancy Lynch. Brewer's conjecture and the feasibility of consistent, available, partition-tolerant web services. *ACM SIGACT News*, 33(2):51–59, 2002.
3. Fay Chang, Jeffrey Dean, Sanjay Ghemawat, Wilson C Hsieh, Deborah A Wallach, Mike Burrows, Tushar Chandra, Andrew Fikes, and Robert E Gruber. Bigtable: A distributed storage system for structured data. *ACM Transactions on Computer Systems (TOCS)*, 26(2):4, 2008.
4. James Manyika, McKinsey Global Institute, Michael Chui, Brad Brown, Jacques Bughin, Richard Dobbs, Charles Roxburgh, and Angela Hung Byers. *Big data: The next frontier for innovation, competition, and productivity*. McKinsey Global Institute, 2011.
5. Rick Cattell. Scalable sql and nosql data stores. *ACM SIGMOD Record*, 39(4):12–27, 2011.
6. Marshall K McKusick and Sean Quinlan. Gfs: Evolution on fast-forward. *ACM Queue*, 7(7):10, 2009.
7. Ronnie Chaiken, Bob Jenkins, Per-Åke Larson, Bill Ramsey, Darren Shakib, Simon Weaver, and Jingren Zhou. Scope: easy and efficient parallel processing of massive data sets. *Proceedings of the VLDB Endowment*, 1(2):1265–1276, 2008.
8. Doug Beaver, Sanjeev Kumar, Harry C Li, Jason Sobel, Peter Vajgel, et al. Finding a needle in haystack: Facebook's photo storage. In *OSDI*, volume 10, pages 1–8, 2010.
9. Giuseppe DeCandia, Deniz Hastorun, Madan Jampani, Gunavardhan Kakulapati, Avinash Lakshman, Alex Pilchin, Swaminathan Sivasubramanian, Peter Vosshall, and Werner Vogels. Dynamo: amazon's highly available key-value store. In *SOSP*, volume 7, pages 205–220, 2007.
10. David Karger, Eric Lehman, Tom Leighton, Rina Panigrahy, Matthew Levine, and Daniel Lewin. Consistent hashing and random trees: Distributed caching protocols for relieving hot spots on the world wide web. In *Proceedings of the twenty-ninth annual ACM symposium on Theory of computing*, pages 654–663. ACM, 1997.
11. Mike Burrows. The chubby lock service for loosely-coupled distributed systems. In *Proceedings of the 7th symposium on Operating systems design and implementation*, pages 335–350. USENIX Association, 2006.
12. Avinash Lakshman and Prashant Malik. Cassandra: structured storage system on a p2p network. In *Proceedings of the 28th ACM symposium on Principles of distributed computing*, pages 5–5. ACM, 2009.
13. Lars George. *HBase: the definitive guide*. O'Reilly Media, Inc., 2011.
14. Doug Judd. hypertable-0.9. 0.4-alpha.
15. Kristina Chodorow. *MongoDB: the definitive guide*. O'Reilly, 2013.
16. Douglas Crockford. The application/json media type for javascript object notation (json). 2006.
17. James Murty. *Programming Amazon Web Services: S3, EC2, SQS, FPS, and SimpleDB*. O'Reilly Media, Inc., 2009.
18. J Chris Anderson, Jan Lehnardt, and Noah Slater. *CouchDB: the definitive guide*. O'Reilly, 2010.
19. Brian F Cooper, Raghu Ramakrishnan, Utkarsh Srivastava, Adam Silberstein, Philip Bohannon, Hans-Arno Jacobsen, Nick Puz, Daniel Weaver, and Ramana Yerneni. Pnuts: Yahoo!'s hosted data serving platform. *Proceedings of the VLDB Endowment*, 1(2):1277–1288, 2008.
20. Brian F Cooper, Adam Silberstein, Erwin Tam, Raghu Ramakrishnan, and Russell Sears. Benchmarking cloud serving systems with ycsb. In *Proceedings of the 1st ACM symposium on Cloud computing*, pages 143–154. ACM, 2010.
21. Tim Kraska, Martin Hentschel, Gustavo Alonso, and Donald Kossmann. Consistency rationing in the cloud: Pay only when it matters. *Proceedings of the VLDB Endowment*, 2(1):253–264, 2009.
22. Kimberly Keeton, Charles B Morrey III, Craig AN Soules, and Alistair Veitch. Lazybase: Freshness vs. performance in information management. *ACM SIGOPS Operating Systems Review*, 44(1):15–19, 2010.

23. Daniela Florescu and Donald Kossmann. Rethinking cost and performance of database systems. *ACM Sigmod Record*, 38(1):43–48, 2009.
24. Maarten Van Steen. Distributed systems principles and paradigms. *Network*, 4:20, 2004.
25. Jeffrey Dean and Sanjay Ghemawat. Mapreduce: simplified data processing on large clusters. *Communications of the ACM*, 51(1):107–113, 2008.
26. Spyros Blanas, Jignesh M Patel, Vuk Ercegovac, Jun Rao, Eugene J Shekita, and Yuanyuan Tian. A comparison of join algorithms for log processing in mapreduce. In *Proceedings of the 2010 ACM SIGMOD International Conference on Management of data*, pages 975–986. ACM, 2010.
27. Hung-Chih Yang and D Stott Parker. Traverse: Simplified indexing on large map-reduce-merge clusters. In *Database Systems for Advanced Applications*, pages 308–322. Springer, 2009.
28. Rob Pike, Sean Dorward, Robert Griesemer, and Sean Quinlan. Interpreting the data: Parallel analysis with sawzall. *Scientific Programming*, 13(4):277–298, 2005.
29. Alan F Gates, Olga Natkovich, Shubham Chopra, Pradeep Kamath, Shravan M Narayanamurthy, Christopher Olston, Benjamin Reed, Santhosh Srinivasan, and Utkarsh Srivastava. Building a high-level dataflow system on top of map-reduce: the pig experience. *Proceedings of the VLDB Endowment*, 2(2):1414–1425, 2009.
30. Ashish Thusoo, Joydeep Sen Sarma, Namit Jain, Zheng Shao, Prasad Chakka, Suresh Anthony, Hao Liu, Pete Wyckoff, and Raghotham Murthy. Hive: a warehousing solution over a map-reduce framework. *Proceedings of the VLDB Endowment*, 2(2):1626–1629, 2009.
31. Michael Isard, Mihai Budiu, Yuan Yu, Andrew Birrell, and Dennis Fetterly. Dryad: distributed data-parallel programs from sequential building blocks. *ACM SIGOPS Operating Systems Review*, 41(3):59–72, 2007.
32. Yuan Yu, Michael Isard, Dennis Fetterly, Mihai Budiu, Úlfar Erlingsson, Pradeep Kumar Gunda, and Jon Currey. Dryadlinq: A system for general-purpose distributed data-parallel computing using a high-level language. In *OSDI*, volume 8, pages 1–14, 2008.
33. Christopher Moretti, Jared Bulosan, Douglas Thain, and Patrick J Flynn. All-pairs: An abstraction for data-intensive cloud computing. In *Parallel and Distributed Processing, 2008. IPDPS 2008. IEEE International Symposium on*, pages 1–11. IEEE, 2008.
34. Grzegorz Malewicz, Matthew H Austern, Aart JC Bik, James C Dehnert, Ilan Horn, Naty Leiser, and Grzegorz Czajkowski. Pregel: a system for large-scale graph processing. In *Proceedings of the 2010 ACM SIGMOD International Conference on Management of data*, pages 135–146. ACM, 2010.
35. Yingyi Bu, Bill Howe, Magdalena Balazinska, and Michael D Ernst. Haloop: Efficient iterative data processing on large clusters. *Proceedings of the VLDB Endowment*, 3(1–2):285–296, 2010.
36. Jaliya Ekanayake, Hui Li, Bingjing Zhang, Thilina Gunarathne, Seung-Hee Bae, Judy Qiu, and Geoffrey Fox. Twister: a runtime for iterative mapreduce. In *Proceedings of the 19th ACM International Symposium on High Performance Distributed Computing*, pages 810–818. ACM, 2010.
37. Matei Zaharia, Mosharaf Chowdhury, Tathagata Das, Ankur Dave, Justin Ma, Murphy McCauley, Michael Franklin, Scott Shenker, and Ion Stoica. Resilient distributed datasets: A fault-tolerant abstraction for in-memory cluster computing. In *Proceedings of the 9th USENIX conference on Networked Systems Design and Implementation*, pages 2–2. USENIX Association, 2012.
38. Pramod Bhatotia, Alexander Wieder, Rodrigo Rodrigues, Umut A Acar, and Rafael Pasquin. Incoop: Mapreduce for incremental computations. In *Proceedings of the 2nd ACM Symposium on Cloud Computing*, page 7. ACM, 2011.
39. Derek G Murray, Malte Schwarzkopf, Christopher Smowton, Steven Smith, Anil Madhavapeddy, and Steven Hand. Ciel: a universal execution engine for distributed data-flow computing. In *Proceedings of the 8th USENIX conference on Networked systems design and implementation*, page 9, 2011.

Chapter 5
Big Data Analysis

Abstract In this chapter, we introduce the methods, architectures and tools for big data analysis. The analysis of big data mainly involves analytical methods for traditional data and big data, analytical architecture for big data, and software used for mining and analysis of big data. Data analysis is the final and the most important phase in the value chain of big data, with the purpose of extracting useful values, providing suggestions or decisions. Different levels of potential values can be generated through the analysis of datasets in different fields.

5.1 Traditional Data Analysis

Traditional data analysis means to use proper statistical methods to analyze massive first-hand data and second-hand data, to concentrate, extract, and refine useful data hidden in a batch of chaotic data, and to identify the inherent law of the subject matter, so as to develop functions of data to the greatest extent and maximize the value of data. Data analysis plays a huge guidance role in making development plans for a country, as well as understanding customer demands and predicting market trend by enterprises.

Big data analysis can be deemed as the analysis of a special kind of data. Therefore, many traditional data analysis methods may still be utilized for big data analysis. Several representative traditional data analysis methods are examined in the following, many of which are from statistics and computer science.

- *Cluster Analysis*: cluster analysis is a statistical method for grouping objects, and specifically, classifying objects according to some features. Cluster analysis is used to differentiate objects with certain features and divide them into some categories (clusters) according to these features,such that objects in the category will have high homogeneity different categories will have high heterogeneity. Cluster analysis is an unsupervised study method without the use of training data.

M. Chen et al., *Big Data: Related Technologies, Challenges and Future Prospects*,
SpringerBriefs in Computer Science, DOI 10.1007/978-3-319-06245-7_5,

- *Factor Analysis*: factor analysis is basically targeted at describing the relation among many indicators or elements with only a few factors, i.e., grouping several closely related variables and then every group of variables becomes a factor (called a factor because it is unobservable, i.e., not a specific variable), and the few factors are then used to reveal the most valuable information of the original data.
- *Correlation Analysis*: correlation analysis is an analytical method for determining the law of correlations among observed phenomena and accordingly conducting forecast and control. There are a plentiful of quantitative relations among observed phenomena such as correlation, correlative dependence, and mutual restriction. Such relations may be classified into two types: (a) function, reflecting the strict dependence relationship among phenomena, which is also called a definitive dependence relationship, among which, every numerical value of a variable corresponds to one or several determined values; (b) correlation, under which some undetermined and inexact dependence relations exist, and a numerical value of a variable may correspond to several numerical values of the other variable, and such numerical values present regular fluctuation surrounding their mean values. A classic example is that customers of many supermarkets purchase beers while they are buying diapers.
- *Regression Analysis*: regression analysis is a mathematical tool for revealing correlations between one variable and several other variables. Based on a group of experiments or observed data, regression analysis identifies dependence relationships among variables hidden by randomness. Regression analysis may change complex and undetermined correlations among variables into simple and regular correlations.
- *A/B Testing*: also called bucket testing. It is a technology for determining plans to improve target variables by comparing the tested group. Big data will require a large number of tests to be executed and analyzed, to ensure sufficient scale of the groups for detecting the significant differences between the control group and the treatment group.
- *Statistical Analysis*: Statistical analysis is based on the statistical theory, a branch of applied mathematics. In statistical theory, randomness and uncertainty are modeled with Probability Theory. Statistical analysis can provide description and inference for large-scale datasets. Descriptive statistical analysis can summarize and describe datasets and inferential statistical analysis draws conclusions from data subject to random variations. Analytical technologies based on complex multi-variate statistical analysis include regression analysis, factor analysis, clustering, and recognition analysis, etc. Statistical analysis is widely applied in the economic and medical care fields [1].
- *Data Mining*: Data mining is a process for extracting hidden, unknown, but potentially useful information and knowledge from massive, incomplete, noisy, fuzzy, and random data. There are also some terms similar to data mining, e.g., discovering knowledge from databases, data analysis, data fusion, and decision support.

Data mining is mainly used to complete the following six different tasks, with corresponding analytical methods: Classification, Estimation, Prediction, Affinity grouping or association rules, Clustering, and Description and Visualization. Original data is deemed as the source to form knowledge and data mining is a process of discovering knowledge from the original data. Original data may be structured data, e.g., data in relational databases, or semi-structured data, e.g., text, graphical, and image data, or even heterogeneous data distributed in the network. Methods to discover knowledge may be mathematical or non-mathematical, and deductive or inductive. Discovered knowledge may be used for information management, query optimization, decision support, and process control, as well as data maintenance.

Mining methods are generally divided into machine learning methods, neural network methods, and database methods. Machine learning may be next divided into inductive learning, example-based learning, and genetic algorithms, etc. Neural network methods may be divided into feedforward neural networks and self-organizing neural networks, etc. Database methods mainly include multi-dimensional data analysis or OLAP (On-Line Analytical Processing), as well as attribute-oriented inductive method.

Various data mining algorithms have been developed, including artificial intelligence, machine learning, mode identification, statistics and database community, etc. In 2006, The IEEE International Conference on Data Mining Series (ICDM) identified ten most influential data mining algorithms through a strict selection procedure [2], including C4.5, k-means, SVM, Apriori, EM, Naive Bayes, and Cart, etc. These ten algorithms cover classification, clustering, regression, statistical learning, association analysis, and linking mining, all of which are the most important problems in data mining research. In addition, other advanced algorithms such as neural networks and genetic algorithms can also be applied to data mining in different applications. Some prominent applications are gaming, business, science, engineering, and supervision, etc.

5.2 Big Data Analytic Methods

In the dawn of the big data era, people are concerned with how to rapidly extract key information from massive data so as to bring values for enterprises and individuals. At present, the main processing methods of big data are shown as follows.

- *Bloom Filter*: Bloom Filter is actually a bit array and a series of Hash functions. The principle of Bloom Filter is to store Hash values of data other than data itself by utilizing a bit array, which is in essence a bitmap index that uses Hash functions to conduct lossy compression storage of data. It has such advantages as high space efficiency and high query speed, but also with some disadvantages like having a certain misrecognition rate and deletion difficulty. Bloom Filter applies to big data applications that allow a certain misrecognition rate.

Table 5.1 Comparison of MPI, MapReduce and Dryad

	MPI	MapReduce	Dryad
Deployment	Computing node and data storage arranged separately (Data should be moved computing node)	Computing and data storage arranged at the same node (Computing should be close to data)	Computing and data storage arranged at the same node (Computing should be close to data)
Resource management/ scheduling	–	Workqueue(google) HOD(Yahoo)	Not clear
Low level programming	MPI API	MapReduce API	Dryad API
High level programming	Null	Pig, Hive, Jaql, ···	Scope, DryadLINQ
Data storage	The local file system, NFS, ···	GFS(google) HDFS(Hadoop), KFS Amazon S3, ···	NTFS, Cosmos DFS
Task partitioning	User manually partition the tasks	Automation	Automation
Communication	Messaging, Remote memory access	Files(Local FS, DFS)	Files, TCP Pipes, Shared-memory FIFOs
Fault-tolerant	Checkpoint	Task re-execute	Task re-execute

- *Hashing*: it is a method that essentially transforms data into shorter fixed-length numerical values or index values. Hashing has such advantages as rapid reading, writing, and high query speed, but a sound Hash function is hard to be found.
- *Index*: index is always an effective method to reduce the expense of disc reading and writing, and improve insertion, deletion, modification, and query speeds in both traditional relational databases that manage structured data, and technologies that manage semi-structured and unstructured data. However, index has a disadvantage that it has the additional cost for storing index files and the index files should be maintained dynamically according to data updates.
- *Triel*: also called trie tree, a variant of Hash Tree. It is mainly applied to rapid retrieval and word frequency statistics. The main idea of Triel is to utilize common prefixes of character strings to reduce comparison on character strings to the greatest extent, so as to improve query efficiency.
- *Parallel Computing*: compared to traditional serial computing, parallel computing refers to utilizing several computing resources to complete a computation task. Its basic idea is to decompose a problem and assign them to several independent processes to be independently completed, so as to achieve co-processing. Presently, some classic parallel computing models include MPI (Message Passing Interface), MapReduce, and Dryad. A qualitative comparison of the three models is presented in Table 5.1.

Although the parallel computing systems or tools, such as MapReduce or Dryad, are useful for big data analysis, they are low levels tools that have a steep learning curve. Therefore, some high-level parallel programming tools or languages are being developed based on these systems. Such high-level languages include Sawzall, Pig, and Hive used for MapReduce, and Scope and DryadLINQ used for Dryad.

5.3 Architecture for Big Data Analysis

Due to the wide range of sources and variety, different structures, and the broad application fields of big data, different analytical architectures shall be considered for big data with different application requirements.

5.3.1 Real-Time vs. Offline Analysis

Big data analysis can be classified into real-time analysis and off-line analysis according to the real-time requirement. Real-time analysis is mainly used in E-commerce and finance. Since data constantly changes, rapid data analysis is needed and analytical results shall be returned with a very short delay. The main existing architectures of real-time analysis include (a) parallel processing clusters using traditional relational databases, and (b) memory-based computing platforms. For example, Greenplum from EMC and HANA from SAP are all real-time analysis architectures.

Offline analysis is usually used for applications without high requirements on response time, e.g., machine learning, statistical analysis, and recommendation algorithms. Offline analysis generally conducts analysis by importing big data of logs into a special platform through data acquisition tools. Under the big data setting, many Internet enterprises utilize the offline analysis architecture based on Hadoop in order to reduce the cost of data format conversion and improve the efficiency of data acquisition. Examples include Facebook's open source tool Scribe, LinkedIn's open source tool Kafka, Taobao's open source tool Timetunnel, and Chukwa of Hadoop, etc. These tools can meet the demands of data acquisition and transmission with hundreds of MB per second.

5.3.2 Analysis at Different Levels

Big data analysis can also be classified into memory level analysis, Business Intelligence (BI) level analysis, and massive level analysis, which are examined in the following.

- *Memory-Level*: Memory-level analysis is for the case when the total data volume is within the maximum level of the memory of a clusters. The memory of current server cluster surpasses hundreds of GB while even the TB level is common. Therefore, an internal database technology may be used and hot data shall reside in the memory so as to improve the analytical efficiency. Memory-level analysis is extremely suitable for real-time analysis. MongoDB is a representative memory-level analytical architecture. With the development of SSD (Solid-State Drive), the capacity and performance of memory-level data analysis has been further improved and widely applied.
- *BI*: BI analysis is for the case when the data scale surpasses the memory level but may be imported into the BI analysis environment. Currently, mainstream BI products are provided with data analysis plans supporting the level over TB.
- *Massive*: Massive analysis for the case when the data scale has completely surpassed the capacities of BI products and traditional relational databases. At present, most massive analysis utilize HDFS of Hadoop to store data and use MapReduce for data analysis. Most massive analysis belongs to the offline analysis category.

5.3.3 Analysis with Different Complexity

The time and space complexity of data analysis algorithms differ greatly from each other according to different kinds of data and application demands. For example, for applications that are amenable to parallel processing, a distributed algorithm may be designed and a parallel processing model may be used for data analysis.

5.4 Tools for Big Data Mining and Analysis

Many tools for big data mining and analysis are available, including professional and amateur software, expensive commercial software, and free open source software. In this section, we briefly review the top five widely used software, according to a survey of "What Analytics, Data mining, Big Data software you used in the past 12 months for a real project" of 798 professionals made by KDNuggets in 2012 [3].

- *R* (30.7 %): R, an open source programming language and software environment, is designed for data mining/analysis and visualization. While compute-intensive tasks are executed, code programmed with C, C++, and Fortran may be in under the R environment. In addition, skilled users may directly call R objects in C. R is a realization of the S language. S is an interpreted language developed by AT&T Bell Labs and used for data exploration, statistical analysis, and drawing plots. Initially, S was mainly implemented in S-PLUS, but S-PLUS is a commercial software. Compared to S, R is more popular since it is open source. R ranks top 1 in the KDNuggets 2012 survey. Furthermore, in a survey of "Design languages you have used for data mining/analysis in the past year" in 2012, R was also in the first place, defeating SQL and Java. Due to the popularity of R, database manufacturers such as Teradata and Oracle both released products supporting R.
- *Excel* (29.8 %): Excel, a core component of Microsoft Office, provides powerful data processing and statistical analysis capability, and aids decision making. When Excel is installed, some advanced plug-ins, such as Analysis ToolPak and Solver Add-in, with powerful functions for data analysis are also integrated but such plug-ins can be used only if users enable them. Excel is also the only commercial software among the top five.
- *Rapid-I Rapidminer* (26.7 %): Rapidminer is an open source software used for data mining, machine learning, and predictive analysis. In an investigation of KDnuggets in 2011, it was more frequently used than R (ranked Top 1). Data mining and machine learning programs provided by RapidMiner include Extract, Transform and Load (ETL), data pre-processing and visualization, modeling, evaluation, and deployment. The data mining flow is described in XML and displayed through a graphic user interface (GUI). RapidMiner is written in Java. It integrates the learner and evaluation method of Weka, and works with R. Functions of Rapidminer are implemented with connection of processes of

operators. The entire flow can be deemed as a production line of a factory, with original data input and model results output. The operators can be regarded as specific functions and feature different input and output characteristics.

- *KNIME* (21.8 %): KNIME (Konstanz Information Miner) is a user-friendly, intelligent, and open-source-rich data integration, data processing, data analysis, and data mining platform [4]. It allows users to create data flows or data channels in a visualized manner, to selectively run some or all analytical procedures, and provides analytical results, models, and interactive views. KNIME was written in Java and, based on Eclipse, provides more functions as plug-ins. Through plug-in files, users can insert processing modules to files, pictures, and time series, and integrate them into various open source projects, e.g., R and Weka. KNIME controls data integration, cleansing, conversion, filtering, statistics, mining, and finally data visualization. The entire development process is conducted under a visualized environment. KNIME is designed as a module-based and expandable framework. There is no dependence between its processing units and data containers, making them adaptive to the distributed environment and independent development. In addition, it is easy to expand KNIME. Developers can effortlessly expand various nodes and views of KNIME.
- *Weka/Pentaho* (14.8 %): Weka, abbreviated from Waikato Environment for Knowledge Analysis, is a free and open-source machine learning and data mining software written in Java. Weka provides such functions as data processing, feature selection, classification, regression, clustering, association rule, and visualization, etc. Pentaho is one of the most popular open-source commercial intelligent software. It is a BI kit based on the Java platform. It includes a web server platform and several tools to support report, analysis, chart, data integration, and data mining, etc., all aspects of BI. Weka's data processing algorithms are also integrated in Pentaho and can be directly called.

References

1. Theodore Wilbur Anderson, Theodore Wilbur Anderson, Theodore Wilbur Anderson, and Theodore Wilbur Anderson. *An introduction to multivariate statistical analysis*, volume 2. Wiley New York, 1958.
2. Xindong Wu, Vipin Kumar, J Ross Quinlan, Joydeep Ghosh, Qiang Yang, Hiroshi Motoda, Geoffrey J McLachlan, Angus Ng, Bing Liu, S Yu Philip, et al. Top 10 algorithms in data mining. *Knowledge and Information Systems*, 14(1):1–37, 2008.
3. What analytics, data mining, big data software you used in the past 12 months for a real project? http://www.kdnuggets.com/polls/2012/analytics-data-mining-big-data-software.html, 2012.
4. Michael R Berthold, Nicolas Cebron, Fabian Dill, Thomas R Gabriel, Tobias Kötter, Thorsten Meinl, Peter Ohl, Christoph Sieb, Kilian Thiel, and Bernd Wiswedel. *KNIME: The Konstanz information miner*. Springer, 2008.

Chapter 6
Big Data Applications

Abstract In the previous chapter, we examined big data analysis, which is the final and most important phase of the value chain of big data. Big data analysis can provide useful values via judgments, recommendations, supports, or decisions. However, data analysis involves a wide range of applications, which frequently change and are extremely complex. In this chapter, the evolution of data sources is reviewed. Then, six of the most important data analysis fields are examined, including structured data analysis, text analysis, website analysis, multimedia analysis, network analysis, and mobile analysis. This chapter is concluded with a discussion of several key application fields of big data.

6.1 Application Evolution

Recently, big data and big data analysis has been proposed for describing datasets and as analytical technologies in large-scale complex programs, which need to be analyzed with advanced analytical methods. As a matter of fact, data driven applications have emerged in the past decades. For example, as early as 1990s, business intelligence has became a prevailing technology for business applications and, network search engines based on massive data mining processing emerged in the early twenty-first century. Some potential and influential applications from different fields and their data and analysis characteristics are discussed as follows.

- *Evolution of Commercial Applications*: The earliest business data was generally structured data, which was collected by companies from old systems and then stored in RDBMSs. Analytical technologies used in such systems were prevailing in 1990s and was intuitive and simple, e.g., reports, instrument panels, special queries, search-based business intelligence, online transaction processing, interactive visualization, score cards, predictive modeling, and data mining [1]. Since the beginning of twenty-first century, networks and websites has been providing a unique opportunity for organizations to have online display and directly interact

M. Chen et al., *Big Data: Related Technologies, Challenges and Future Prospects*,
SpringerBriefs in Computer Science, DOI 10.1007/978-3-319-06245-7_6,

with customers. Abundant products and customer information, including click stream data logs and user behavior, etc., can be acquired from the websites. Product layout optimization, customer trade analysis, product suggestions, and market structure analysis can be conducted by text analysis and website mining technologies. As reported in [2], the quantity of mobile phones and tablet PC first surpassed that of laptops and PCs in 2011. Mobile phones and Internet of Things based on sensors are opening a new generation of innovation applications, and searching for larger capacity of supporting location sensing, people oriented, and context operation.

- *Evolution of Network Applications*: The early Internet mainly provided email and webpage services. Text analysis, data mining, and webpage analysis technologies have been applied to the mining of email contents and building search engines. Nowadays, most applications are web-based, regardless of their application field and design goals. Network data accounts for a major percentage of the global data volume. Web has became a common platform for interconnected pages, full of various kinds of data, such as text, images, videos, pictures, and interactive contents, etc. Therefore, plentiful advanced technologies used for semi-structured or unstructured data emerged at the right moment. For example, the image analysis technology may extract useful information from pictures, e.g., face recognition. Multimedia analysis technologies can be applied to the automated video surveillance systems for business, law enforcement, and military applications. Since 2004, online social media, such as Internet forums, online communities, blogs, social networking services, and social multimedia websites, etc., provide users with great opportunities to create, upload, and share contents generated by users. Different user groups may search for daily news and celebrity news, publish their social and political opinions, and provide different applications with timely feedback.
- *Evolution of Scientific Applications*: Scientific research in many fields is acquiring massive data with high-throughput sensors and instruments, such as astrophysics, oceanology, genomics, and environmental research. The U.S. National Science Foundation (NSF) has recently announced the BIGDATA Research Initiative to promote research efforts to extract knowledge and insights from large and complex collections of digital data. Some scientific research disciplines have developed massive data platforms and obtained useful outcomes. For example, in biology, iPlant [3] applies network infrastructure, physical computing resources, coordination environment, virtual machine resources, and inter-operative analysis software and data service to assist researches, educators, and students in enriching all plant sciences. IPlant dataset have high varieties in form, including specification or reference data, experimental data, analog or model data, observation data, and other derived data.

6.2 Big Data Analysis Fields

Data analysis research can be divided into six key technical fields, i.e., structured data analysis, text data analysis, website data analysis, multimedia data analysis, network data analysis, and mobile data analysis. Such a classification aims to emphasize data characteristics, but some of the fields may utilize similar technologies. Since data analysis has a broad scope and it is not easy to have a comprehensive coverage, we will focus on the key problems and technologies in data analysis in the following discussions.

6.2.1 Structured Data Analysis

Business applications and scientific research may generate massive structured data, of which the management and analysis rely on mature commercialized technologies, such as RDBMS, data warehouse, OLAP, and BPM (Business Process Management) [4]. Data analysis is mainly based on data mining and statistical analysis, both of which have been well studied over the past 30 years.

Data analysis is still a very active research field and new application demands drive the development of new methods. Statistical machine learning based on exact mathematical models and powerful algorithms have been applied to anomaly detection [5] and energy control [6]. Exploiting data characteristics, time and space mining may extract knowledge structures hidden in high-speed data flows and sensor data models and modes [7]. Driven by privacy protection in e-commerce, e-government, and health care applications, privacy protection data mining is an emerging research field [8]. Over the past decade, benefited by the substantial popularization of event data, new process discovery, and consistency check technologies, process mining is becoming a new research field especially in process analysis with event data [9].

6.2.2 Text Data Analysis

The most common format of information storage is text, e.g., email communication, business documents, web pages, and social media. Therefore, text analysis is deemed to feature more business-based potential than structured data mining. Generally, tax analysis, also called text mining, is a process to extract useful information and knowledge from unstructured text. Text mining is an inter-disciplinary problem, involving information retrieval, machine learning, statistics, computing linguistics, and data mining in particular. Most text mining systems are based on text expressions and natural language processing (NLP), with more focus on the latter.

Document introduction and query processing are the foundation for developing vector space model, Boolean Retrieval Model, and probability retrieval model, which constitute the foundation of search engines. Since the early 1990s, search engines have evolved into a mature business system, which generally consist of rapidly distributed crawling, effectively inverted index, webpage sequencing based on inlink, and search log analysis [10].

NLP can enable computers to analyze, interpret, and even generate text. Some common NLP methods are: lexical acquisition, word sense disambiguation, part-of-speech tagging, and probabilistic context free grammar [11]. Some NLP-based technologies have been applied to text mining, including information extraction, topic models, text summarization, classification, clustering, question answering, and opinion mining. Information mining shall automatically extract specific structured information from texts. Named entity recognition (NER) technology, as a subtask of information extraction, aims to recognize atomic entities in texts subordinate to scheduled categories (e.g. figures, places, and organizations), which have been successfully applied to the development of new analysis [12] and medical applications [13] recently. The topic models are built according to the opinion that "documents are constituted by topics and topics are the probability distribution of vocabulary." Topic models are models generated by documents, stipulating the probability program to generate documents.

Presently, various probabilistic topic models have been used to analyze document contents and lexical meanings [14]. Text summarization is to generate a reduced summary or extract from a single or several input text files. Text summarization may be classified into concrete summarization and abstract summarization [15]. Concrete summarization selects important sentences and paragraphs from source documents and concentrates them into shorter forms. Abstract summarization may interpret the source texts and, according to linguistic methods, use a few words and phrases to represent the source texts.

Text classification is to recognize probabilistic topic of documents by putting documents in scheduled topics. Text classification based on the new graph representation and graph mining has recently attracted considerable interest [16]. Text clustering is used to group similar documents with scheduled topics, which is different from text classification that gathers documents together. In text clustering, documents may appear in multiple subtopics. Generally, some clustering algorithms in data mining can be utilized to compute the similarities of documents. However, it is also shown that the structural relationship information may be exploited to improve the clustering performance in Wikipedia [17]. The question answering system is designed to search for the optimal answer to a given question. It involves different technologies of question analysis, source retrieval, answer extraction, and answering demonstration [18]. The question answering system may be applied in many fields, including education, website, healthcare, and national defense. Opinion mining, similar to sentiment analysis, refers to the computing technologies for identifying and extracting subjective information from news assessment, comment, and other user-generated contents. It provides opportunities for users to understand the opinions of the public and customers on social events, political movements, business strategies, marketing activities, and product preference [19].

6.2.3 Web Data Analysis

Over the past decade, we have witnessed the explosive growth of Internet information. Web analysis has emerged as an active research field. Web analysis aims to automatically retrieve, extract, and evaluate information from Web documents and services so as to discover useful knowledge. Web analysis is related to several research fields, including database, information retrieval, NLP, and text mining. According to the different parts of the Web to be mined, we classify Web analysis into three related fields: Web content mining, Web structure mining, and Web usage mining [20].

Web content mining is the process to discover useful knowledge in Web pages, which generally involve several types of data, such as text, image, audio, video, code, metadata, and hyperlink.

The research on image, audio, and video mining has recently been called multimedia analysis, which will be discussed in Sect. 6.2.4. Since most Web content data is unstructured text data, the research on Web data analysis mainly centers around text and hypertext. Text mining is discussed in Sect. 6.2.2, while Hypertext mining involves mining semi-structured HTML files that contain hyperlinks.

Supervised learning and classification play important roles in hyperlink mining, e.g., email, newsgroup management, and Web catalogue maintenance [21]. Web content mining can be conducted with two methods: the information retrieval method and the database method. Information retrieval mainly assists in or improves information lookup, or filters user information according to deductions or configuration documents. The database method aims to simulate and integrate data in Web, so as to conduct more complex queries than searches based on key words.

Web structure mining involves models for discovering Web link structures. Here, the structure refers to the schematic diagrams linked in a website or among multiple websites. Models are built based on topological structures provided with hyperlinks with or without link description. Such models reveal the similarities and correlations among different websites and are used to classify website pages. Page Rank [22] and CLEVER [23] make full use of the models to look up related website pages. Topic-oriented crawler is another successful case by utilizing the models [24]. Topic-oriented crawler is targeted at selectively discovering pages related to scheduled topic sets. Top-oriented crawler may analyze crawling boundary to look for links mostly related to crawling and to avoid the involvement of irrelevant areas, other than collecting and indexing all accessible webpage files, so as to answer all possible Ad-Hoc queries. This way, a great quantity of hardware and network resources may be saved and crawling updating task may be assisted.

Web usage mining aims to mine auxiliary data generated by Web dialogues or behaviors. Web content mining and Web structure mining use the master Web data. Web usage data includes access logs at Web servers, logs at proxy servers, browsers' history records, user profiles, registration data, user sessions or trades, cache, user queries, bookmark data, mouse click and scroll, and any other kind of data generated through interaction with the Web. As Web services and the Web2.0 are becoming mature and popular, Web usage data will have increasingly high variety. Web usage mining plays key roles in personalized space, e-commerce, network privacy/security, and other emerging fields. For example, collaborative recommender systems can personalize e-commerce by utilizing the different preferences of users.

6.2.4 Multimedia Data Analysis

Multimedia data (mainly including images, audios, and videos) have been growing at an amazing speed. Multimedia content sharing is to extract related knowledge and understand semantemes contained in multimedia data. Because multimedia data is heterogeneous and most of such data contains richer information than simple structured data and text data, extracting information is confronted with the huge challenge of the semantic differences of multimedia data. Research on multimedia analysis covers many disciplines. Some recent research priorities include multimedia summarization, multimedia annotation, multimedia index and retrieval, multimedia suggestion, and multimedia event detection, etc.

Audio summarization can be accomplished by simply extracting the prominent words or phrases from metadata or synthesizing a new representation. Video summarization is to interpret the most important or representative video content sequence, and it can be static or dynamic. Static video summarization methods utilize a key frame sequence or context-sensitive key frames to represent a video. Such methods are very simple and have been applied to many business applications (e.g., Yahoo!, Alta Visa, and Google), but the playback performance is poor. Dynamic summarization methods use a series of video clips to represent a video, configure low-level video functions, and take other smooth measures to make the final summarization look more natural. In [25], the authors proposed a topic-oriented multimedia summarization system (TOMS) that can automatically summarize the important information in a video belonging to a certain topic area, based on a given set of extracted features from the video.

Multimedia annotation inserts labels to describe contents of images and videos in both syntax and semantic levels. With the assistance of such labels, the management, summarization, and retrieval of multimedia data can be easily implemented. Since manual annotation is both time and labor intensive, multimedia automatic annotation without any human interventions becomes highly appealing. The main challenge for multimedia automatic annotation is semantic difference, i.e. the difference between low-level features and annotations. Although much progress has been made, the performance of the existing automatic annotation methods still needs to be improved. Currently, many efforts are being made to synchronously explore both manual and automatic multimedia annotation [26].

Multimedia index and retrieval involve describing, storing, and organizing multimedia information and assisting users to conveniently and quickly look up multimedia resources [27]. Generally, multimedia index and retrieval include five procedures: structural analysis, feature extraction, data mining, classification and annotation, query and retrieval [28]. Structural analysis aims to segment a video into several semantic structural elements, including lens boundary detection, key frame extraction, and scene segmentation, etc. According to the result of structural analysis, the second procedure is feature extraction, which mainly includes further mining the features of necessary key frames, objects, texts, and movements, which are the foundation of video index and retrieval. Data mining, classification, and annotation are generated to utilize the extracted features to find the modes of video contents and put videos into scheduled categories so as to generate video indexes. Upon receiving a query, the system will use a similarity measurement method to look up a candidate video. The retrieval result optimizes the related feedback.

Multimedia recommendation aims to recommend specific multimedia contents according to users' preferences. It is proven to be an effective approach to provide quality personalized services. Most existing recommendation systems can be classified into content-based systems and collaborative-filtering-based systems. The content-based methods identify users or general features in which the users are interested, and recommend users for other contents with similar features. These methods purely rely on content similarity measurement but most of them are limited by content analysis and excessive specifications. The collaborative-filtering-based methods identify groups with similar interests and recommend contents for group members according to their behaviors [29]. Presently, a mixed method is introduced, which integrates advantages of the aforementioned two types of methods to improve the recommendation quality [30].

The U.S. NIST initiated the TREC Video Retrieval Evaluation detecting the occurrence of an event in video-clips based on Event Kit, which contains some text description related to concepts and video examples [31]. The research on video event detection is still in its infancy. The existing research on event detection mainly focuses on sports or news events, running or abnormal events in monitoring videos, and other similar events with repetitive patterns. In [32], the author proposed a new algorithm on special multimedia event detection using a few positive training examples.

6.2.5 Network Data Analysis

Network analysis evolved from the initial quantitative analysis [33] and sociological network analysis [34] into the emerging online social network analysis in the beginning of twenty-first century. Many prevailing online social networking services include Twitter, Facebook, and LinkedIn, etc. have been increasingly popular over the years. Such online social networking services generally include massive linked data and content data. The linked data is mainly in the form of graphic structures, describing the communications between two entities. The content data contains text, image, and other network multimedia data. The rich contents of such networks bring about both unprecedented challenges and opportunities to data analysis. In accordance with the data-centered perspective, the existing research on social networking service contexts can be classified into two categories: link-based structural analysis and content-based analysis [35].

The research on link-based structural analysis has always been committed on link prediction, community discovery, social network evolution, and social influence analysis, etc. SNS may be visualized as graphs, in which every vertex corresponds to a user and edges correspond to the correlations among users. Since SNS are dynamic networks, new vertexes and edges are continually added to the graphs. Link prediction is to predict the possibility of future connection between two vertexes. Many technologies can be used for link prediction, e.g., feature-based classification, probabilistic methods, and Linear Algebra. Feature-based classification is to select a group of features for a vertex and utilize the existing link information to generate binary classifiers to predict the future link [36]. Probabilistic methods aim to build models for connection probabilities among vertexes in SNS [37]. Linear Algebra computes the similarity between two vertexes according to the singular similar matrix [38]. A community is represented by a sub-graphic matrix, in which edges connecting vertexes in the sub-graph feature high density, while the edges between two sub-graphs feature much lower density [39].

Many methods against community detection have been proposed and studied, most of which are topology-based target functions relying on the concept of capturing community structure. Du et al. utilized the property of overlapping communities in real life to propose a more effective large-scale SNS community detection method [40]. The research on SNS aims to look for a law and deduction model to interpret network evolution. Some empirical studies found that proximity bias, geographical limitations, and other factors play important roles in SNS evolution [41–43], and some generation methods are proposed to assist network and system design [44].

Social influence refers to the case when individuals change their behavior under the influence of others. The strength of social influence depends on the relation among individuals, network distances, time effect, and characteristics of networks and individuals, etc. Marketing, advertisement, recommendation, and other applications can benefit from social influence by qualitatively and quantitatively measuring the influence of individuals on others [45, 46]. Generally, if the proliferation of contents between SNS are considered, the performance of link-based structural analysis may be further improved.

Benefited by the revolutionary progress of Web2.0, the use of generated contents is explosively growing in SNS. SNS is used to generated contents by various technology, including blogs, micro blogs, opinion mining, photos, video sharing, social bookmarking, social network sites, social news, and Wiki. Content-based analysis in SNS is also known as social media analysis. Social media include text, multimedia, positioning, and comments. Nearly all research topics related to structural analysis, text analysis, and multimedia analysis may be interpreted as social media analysis, but social media analysis is confronted with unprecedented challenges. First, massive and continually growing social media data should be automatically analyzed within a reasonable time. Second, social media data contains much noise, e.g., blogosphere contains a large number of spam blogs, and so does trivial Tweets in Twitter. Third, SNS are dynamic networks, which are frequently and quickly changed and updated.

Since social media is close to SNS, social media analysis is inevitably influenced by SNS analysis. SNS analysis refers to the text analysis of SNS context and characteristics of social and network structures, as well as multimedia analysis. The existing research on social media analysis is still in its infancy. The applications of SNS text analysis include transfer learning in keyword search, classification, clustering, and heterogeneous networks. Keyword search tries to synchronously use contents and link behaviors for search [47]. The motivation for such applications is that text files containing similar keywords are generally connected to each other [48]. During classification, assuming all nodes of the SNS are provided with labels, the nodes added with labels are classified. During clustering, researchers aim to determine node sets with similar contents and accordingly group them [49]. Considering that SNS contains massive information of different interlinked objects, e.g., articles, labels, images, and videos, transfer learning in heterogeneous networks aims to transfer knowledge information among different links [50].

Multimedia datasets in SNS is organized in a structured form, which brings rich information, e.g., semantic ontology, social interaction, community media, geographical maps, and multimedia opinions. Structural multimedia analysis in SNS is also called multimedia information networks. The link structure of multimedia information networks is mainly a logic structure, which are of vital importance to the multimedia in multimedia networks. The logic connection structures in multimedia information networks can be classified into four types: semantic ontology, community media, individual photo albums, and geographical positions [36].

6.2.6 Mobile Traffic Analysis

With the rapid growth of mobile computing, mobile terminals and applications in the world are growing rapidly. By April 2013, Android Apps has provided more than 650,000 applications, covering nearly all categories. By the end of 2012, the monthly mobile data flow has reached 885 PB [51]. The massive data and abundant applications exploit a broad research field for mobile analysis but also bring about a few challenges. As a whole, mobile data has unique characteristics, e.g., mobile sensing, moving flexibility, noise, and a large amount of redundancy. Recently, new research on mobile analysis has been started in different fields. Because of the far immaturity of the research on mobile analysis, we will only introduce some recent and representative analysis applications in this section.

With the growth of numbers of mobile users and the improved performance, mobile phones are now useful for building and maintaining communities, such as communities based on geographical locations and communities based on different cultures and interests, e.g., the latest Wechat. Traditional network communities or SNS communities are in short of online interaction among members, and the communities are active only when members are sitting before computers. On the contrary, mobile phones can support rich interaction any time and anywhere. Wechat supports not only one-to-one communications, but also many-to-many communication. Mobile communities are defined as that a group of individuals with the same hobbies (i.e., health, safety, and entertainment, etc.) gather together on networks, meet to make a common goal, decide measures through consultation to achieve the goal, and start to implement their plan [52]. In [53], the authors proposed a qualitative model of a mobile community. It is now widely believed that mobile community applications will greatly promote the development of the mobile industry.

RFID labels are used to identify, locate, track, and supervise physical objects in a cost-effective manner. RFID is widely applied to inventory management and logistics. However, RFID brings about many challenges to data analysis: (a) RFID data is very noisy and redundant; (b) RFID data is instant and streaming data with a huge volume and limited processing time. We can track objects and monitor system status by deducing some original events through mining the semantics of RFID data, including location, cluster, and time, etc. In addition, we may design the application logic as complex events and then detect such complex events, so as to realize more advanced business applications. In [54], the authors discussed a shoplifting case as an advanced complex event.

Recently, the progress in wireless sensor, mobile communication technology, and stream processing enable people to build a body area network to have real-time monitoring of people's health. Generally, medical data from different sensors has different characteristics, e.g., heterogeneous attribute sets, different time and space relations, and different physiological relations, etc. In addition, such datasets involve privacy and safety protection. In [55], Garg and others introduced a multi-modal transport analysis mechanism of raw data for real-time monitoring of health. Under the circumstance that only highly comprehensive characteristics related to health are available, Park et al. in [56] examined approaches to better utilize such comprehensive information to strength data at all levels. Comprehensive statistics of some partitions is used to recognize clustering and input a characteristic value with a more comprehensive degree. The input characteristics will be further used to predict modeling so as to improve performance.

Researchers from Gjovik University College in Norway and Derawi Biometrics united to develop an application for smart phones, which analyzes paces when people walk and uses the paces for unlocking the safety system [57]. In the meanwhile, Robert Delano and Brian Parise from Georgia Institute of Technology developed an application called iTrem, which monitors human bodies' trembling with a built-in seismograph in a mobile phone, so as to cope with Parkinson and other nervous system diseases [57]. Many other mobile device applications aim to acquire information through mobile devices, no matter how useful such information is for future data analysis.

6.3 Key Applications

6.3.1 Application of Big Data in Enterprises

At present, big data mainly comes from and used in enterprises, while BI and OLAP can be regarded as the predecessors of big data application. The application of big data in enterprises can enhance their production efficiency and competitiveness in many aspects. In particular, on marketing, with correlation analysis of big data, enterprises can more accurately predict the behavior of consumers and mine new business modes. On sales planning, after comparison of massive data, enterprises can optimize their commodity prices. On operation, enterprises can improve their operation efficiency and operation satisfaction, optimize the input of labor force, accurately forecast personnel allocation requirements, avoid excess production capacity, and reduce labor cost. On supply chain, using big data, enterprises may conduct inventory optimization, logistic optimization, and supplier coordination, etc., to mitigate the gap between supply and demand, control budgets, and improve services.

In finance, the application of big data in enterprises has been rapidly developed. For example, China Merchants Bank (CMB) utilizes data analysis to recognize that such activities as “Multi-times score accumulation” and “score exchange in shops,” are effective for attracting quality customers. By building a customer loss early warning model, the bank can sell high-yield financial products to the top 20 % customers in loss ratio so as to retain them. As a result, the loss ratios of customers with Gold Cards and Sunflower Cards have been reduced by 15 % and 7 %, respectively. By analyzing customers’ transaction records, potential small and micro corporate customers can be effectively identified. By utilizing remote banking and the cloud referral platform to implement cross-selling, considerable performance gains were achieved.

Obviously, the most classic application is in e-commerce. Tens of thousands of transactions are conducted in Taobao and the corresponding transaction time, commodity prices, and purchase quantities are recorded every day. More important, such information matches age, gender, address, and even hobbies and interests of buyers and sellers. Data Cube of Taobao is a big data application on the Taobao platform, through which, merchants can be ware of the macroscopic industrial status of the Taobao platform, market conditions of their brands, and consumers’ behaviors, etc., and accordingly make production and inventory decisions. Meanwhile, more consumers can purchase their favorite commodities with more preferable prices.

The credit loan of Alibaba automatically analyzes and judges if to lend loans to enterprises through the acquired enterprise transaction data by virtue of big data technologies, while manual intervention does not occur in the entire process. It is disclosed that, so far, Alibaba has lent more than RMB 30 billion Yuan, with the rate of bad loans of only about 0.3 %, which greatly lower than those of other commercial banks.

6.3.2 Application of IoT Based Big Data

Internet of Things is not only an important source of big data, but also the main market of application of big data. In Internet of Things, every object in the real world may be both the producer and consumer of data and, because of the high variety of objects, the applications of Internet of Things also evolve endlessly.

Logistic enterprises may have profoundly experienced with the application of big data of Internet of Things. Trucks of UPS are installed with sensors, wireless adapters, and GPS, so the Headquarter can track truck positions and prevent engine failures. Meanwhile, this equipment also help UPS supervise and manage its employees, and optimize delivery routes. The optimal delivery routes specified by UPS for trucks are derived from their past driving experience. In 2011, UPS drivers have driven for nearly 48.28 million km less.

Smart city is a hot research area based on the application of Internet of Things data. The U.S. Miami-Dade County is a sample of smart city. The smart city project cooperation between Miami-Dade County in Florida and IBM closely connects 35 types of key county government departments and Miami City, and helps government leaders obtain better information support in decision making for managing water resources, reducing traffic jam, and improving public safety. IBM provides Dade County with smart instrument panel application by virtue of the in-depth analysis under cloud computing, so as to help the departments of county government with coordination-based and visualized management. The application of smart city brings about benefits in many aspects for Dade County. For example, Department of Park Management of Dade County saved one million USD in water bills due to timely identifying and fixing water pipes that were running and leaking this year.

6.3.3 Application of Online Social Network-Oriented Big Data

Online SNS is a social structure constituted by social individuals and connections among individuals based on an information network. Big data of online SNS mainly comes from instant messages, online social, micro blog, and shared space, etc. Since the big data of online SNS represents various user activities, the analysis of such data receives more attention. The analysis of big data of online SNS uses computational analytical method provided for understanding relations in the human society by virtue of theories and methods, which involves mathematics, informatics, sociology, and management science, etc., from three dimensions including network structure, group interaction, and information spreading. The application of big data of online SNS includes network public opinion analysis, network intelligence collection and analysis, socialized marketing, government decision-making support, and online education, etc. Figure 6.1 illustrates the technical framework of the application of big data of online SNS. Classic applications of big data of online SNS are introduced in the following, which mainly mine and analyze content information and structural information to acquire values.

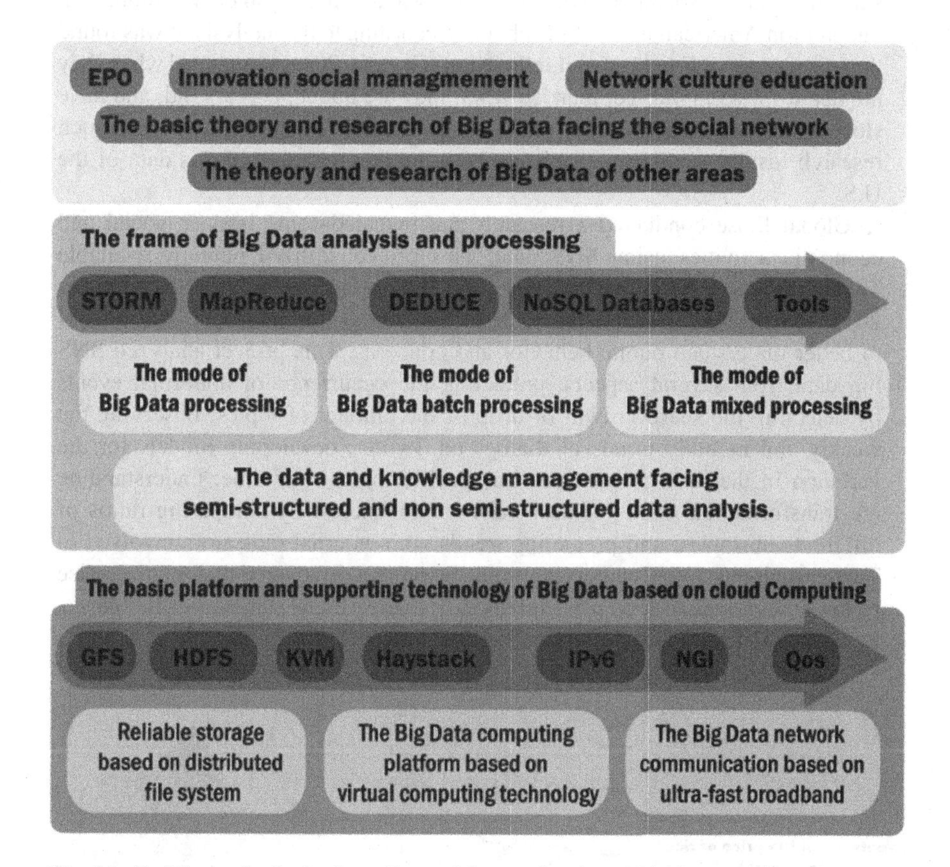

Fig. 6.1 Enabling technologies for online social network-oriented big data

- *Content-Based Applications*: Language and text are two most important forms of representation in SNS. Through the analysis of language and text, user preferences, emotions, interests, and demands, etc. may be revealed.
- *Structure-Based Applications*: On SNS with users as nodes, social relation, interest, and hobbies, etc. aggregate relations among users into a clustered

structure. Such structure with close relations among internal individuals but loose externally relations is also called a community. The community-based analysis is of vital importance to improve information propagation and for the research on interpersonal relation analysis.

The U.S. Santa Cruz Police Department experimented by applying data to conducting predictive analysis. By analyzing SNS, the police department can discover crime trends and crime modes, and even predict the crime rates in major regions [57].

In April 2013, Wolfram Alpha, a computing and search engine of the U.S., studied the law of social behaviors of users by analyzing social data of more than one million American users of Facebook. According to the analysis, it was found that most users of Facebook fall in love in their early 20s, get engaged when they are about 27 years old, get married when they are about 30 years old, and have slow changes in their marriage relationship between 30 and 60 years old. Such research results are highly consistent with the demographic census data of the U.S.

Global Pulse conducted a research that revealed some laws in social and economic activities using SNS data. This project utilized publicly available Twitter messages in English, Japanese, and Indonesian from July 2010 to October 2011, to analyze topics related to food, fuel, housing, and loan. The goal is to better understand public behavior and concerns. This project analyzed SNS big data from several aspects: predicting the occurrence of abnormal events by detecting the sharp growth or drop of the amount of topics; observing the weekly and monthly trends of dialogs on Twitter; developing models for the variation in the level of attention on specific topics over time; understanding the transformation trends of user behavior or interest by comparing ratios of different sub-topics; and predicting trends with external indicators involved in Twitter dialogues. As a classic example, the project discovered that the rice price follows the change of food price inflation from the official statistics of Indonesia, by analyzing topics related to rice price on Twitter (Fig. 6.2).

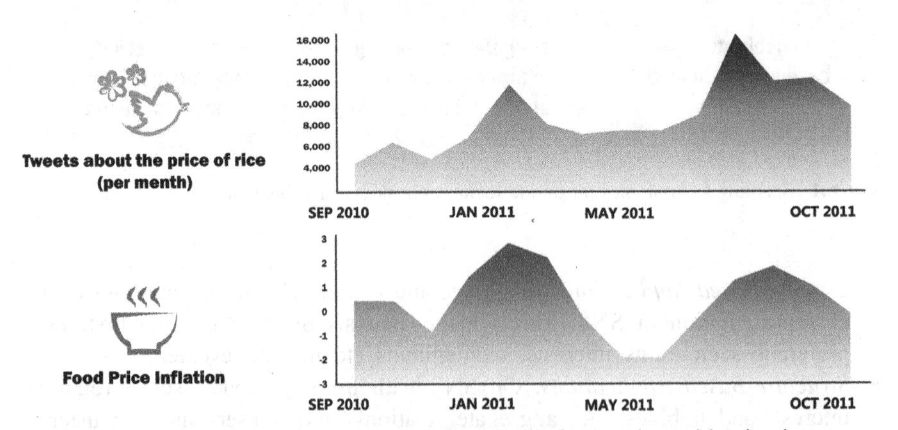

Fig. 6.2 The correlation between Tweets about the price of rice and food price inflation

Generally speaking, the application of big data of online SNS may help to better understand people's behavior and master the laws of social and economic activities from the following three aspects:

- *Early Warning*: to rapidly cope with crisis if any by detecting abnormalities in the usage of electronic devices and services.
- *Real-Time Monitoring*: to provide accurate information for the formulation of polices and plans by monitoring the current behavior, emotion, and preference of users.
- *Real-Time Feedback*: acquire groups' feedbacks against some social activities based on real-time monitoring.

The application of big data of online SNS involves three core technical problems:

- *Data Model*: Most traditional SNS data models are based on the static mode and specific analytical algorithms, and are not amenable for effective computation with data in the PB and higher scales. On the other hand, SNS analysis usually implements multi-dimensional complex relevant analysis on dynamic data. New theories and models need to be investigated to bridge this gap.
- *Data Storage and Management*: The existing Internet based storage management methods mainly support big data storage and rapid query. However, the existing approach does not effectively support the analytical computation of big data of online SNS, featuring high correlation, dynamic variability, and multi-dimensional evolution, etc. Therefore, new storage and management methods need to be developed.
- *Data Analysis*: The existing analytical methods on big data of SNS are mainly based on single-dimensional attribute, with insufficient accuracy. On the other hand, SNS analysis, such as topic evolution, group interaction, and public emotion drifting, etc., usually incorporates complex correlation analysis from the perspective of structure, group, and information. There is a need for the basic theory and methods to support complex correlation, multi-dimensional, large-scale, dynamic data.

6.3.4 Applications of Healthcare and Medical Big Data

Medical data is continuously and rapidly growing containing abundant and various information values. Big data has unlimited potential for effectively storing, processing, querying, and analyzing medical data. The application of medical big data will profoundly influence the human health.

For example, Aetna Life Insurance Company selected 102 patients from a pool of a 1,000 patients to complete an experiment in order to help predict the recovery of patients with metabolic syndrome. In an independent experiment, it scanned 600,000 laboratory test results and 180,000 claims through a series of detection test results of metabolic syndrome of patients in three consecutive years. In addition, it summarized the final result into an extreme personalized treatment plan to assess the dangerous factors and main treatment plans of patients. This way, doctors may reduce morbidity by 50 % in the next 10 years by prescribing statins and helping patients to lose weight by five pounds, or suggesting patients to reduce the total triglyceride in their bodies if the sugar content in their bodies is over 20 %.

The Mount Sinai Medical Center in the U.S. utilizes technologies of Ayasdi, a big data company, to analyze all genetic sequences of Escherichia Coli, including over one million DNA variants, to know why bacterial strains resist antibiotics. Ayasdi's technology uses Topological data analysis, a brand-new mathematic research method, to understand data characteristics. HealthVault of Microsoft is an excellent application of medical big data launched in 2007. The goal is to manage individual health information in individual and family medical equipment. Presently, health information can be entered and uploaded with mobile smart devices and imported into individual medical records by a third-party agency. In addition, it can be integrated with a third-party application with the software development kit (SDK) and open interface.

6.3.5 Collective Intelligence

With the rapid development of wireless communication and sensor technologies, mobile phones and tablet computers have integrated more and more sensors, with increasingly stronger computing and sensing capacities. As a result, crowd sensing is coming to the center stage of mobile computing. In crowd sensing, a large number of general users utilize mobile devices as basic sensing units to conduct coordination with mobile networks for distribution of sensed tasks and collection and utilization of sensed data. The goal is to complete large-scale and complex social sensing tasks. In crowd sensing, participants who complete complex sensing tasks do not need to have professional skills. Crowd sensing modes represented by Crowdsourcing has been successfully applied to geotagged photograph, positioning and navigation, urban road traffic sensing, market forecast, opinion mining, and other labor-intensive applications.

Crowdsourcing, a new approach for problem solving, takes a large number of general users as the foundation and distributes tasks in a free and voluntary way. Crowdsourcing can be useful for labor-intensive applications, such as picture marking, language translation, and speech recognition. The main idea of Crowdsourcing is to distribute tasks to general users and to complete tasks that users could not individually complete or do not anticipate to complete. With no need for intentionally deploying sensing modules and employing professionals, Crowdsourcing can broaden the sensing scope of a sensing system to reach the city scale and even larger scales.

As a matter of fact, Crowdsourcing has been applied by many companies before the emergence of big data. For example, P & G, BMW, and Audi improve their R & D and design capacities by virtue of Crowdsourcing. In the big data era, Spatial Crowdsourcing becomes a hot topic. The operation framework of Spatial Crowdsourcing is shown as follows. A user may request the service and resources related to a specified location. Then the mobile users who are willing to participate in the task will move to the specified location to acquire related data (such as video, audio, or pictures). Finally, the acquired data will be send to the service requester. With the rapid growth of usage of mobile devices and the increasingly complex functions provided by mobile devices, it can be forecasted that Spatial Crowdsourcing will be more prevailing than traditional Crowdsourcing, e.g., Amazon Turk and Crowdflower.

6.3.6 Smart Grid

Smart Grid is the next generation power grid constituted by traditional energy networks integrated with computation, communications and control for optimized generation, supply, and consumption of electric energy. Smart Grid related big data are generated from various sources, such as (a) power utilization habits of users, (b) phasor measurement data, which are measured by phasor measurement unit (PMU) deployed national-wide, (c) energy consumption data measured by the smart meters in the Advanced Metering Infrastructure (AMI), (d) energy market pricing and bidding data, (e) management, control and maintenance data for the devices and equipment in the power generation, transmission and distribution networks (such as Circuit Breaker Monitors and transformers). Smart Grid brings about the following challenges on exploiting big data.

- *Grid Planning*: By analyzing data in Smart Grid, the regions can be identified that have excessive high electrical load or power outage frequencies. Even the transmission lines with high failure possibility can be predicted. Such analytical results may contribute to grid upgrading, transformation, and maintenance, etc. For example, researchers from University of California, Los Angeles designed an "electric map" according to the big data theory and made a California map by integrating census information and real-time power utilization information provided by electric power companies. The map takes a block as a unit to demonstrate the power consumption of every block at the moment. It can even compare the power consumption of the block with the average income per capita and building types, so as to obtain more accurate power usage habits of all kinds of groups in the community. This map provides effective and visual load forecast for city and power grid planning. Preferential transformation on power grid facilities in blocks with high power outage frequencies and serious overloads may be conducted, as displayed in the map.
- *Interaction Between Power Generation and Power Consumption*: An ideal power grid shall balance power generation and power consumption. However, the traditional power grid is constructed based on one-directional approach of

transmission-transformation-distribution-consumption, which could not adjust the generation capacity according to the demand of power consumption, thus leading to electric energy redundancy and waste. To this end, smart electric meters are developed to enable the interaction between power consumption and power generation, and to improve power supply efficiency. TXU Energy has widely deployed smart electric meters with a big success. Power supply companies can read power utilization data every other 15 min other than every month in the past. Therefore, labor cost for meter reading is saved and, because power utilization data (a source of big data) are frequently and rapidly acquired and analyzed, power supply companies can adjust the electricity price according to peak and low periods of power consumption. TXU Energy utilized such price lever to stabilize the peak and low fluctuations of power consumption. As a matter of fact, the application of big data in the smart grid can help the realization of time-sharing dynamic pricing, which is a win-win situation for both energy suppliers and users.

- *Access of Intermittent Renewable Energy*: At present, many new energy resources, such as wind energy and solar energy, are also accessed to power grids. However, since the power generation capacities of such new energy resources are closely related to climate conditions that feature randomness and intermittency, it is challenging to access them to power grids. If the big data of power grids is effectively analyzed, such intermittent renewable new energy sources can be effectively managed: the electricity generated by the new energy resources can be allocated to regions with electricity shortage. Such energy resources can complement the traditional hydropower and thermal power generations.

References

1. Rita L Sallam, James Richardson, John Hagerty, and Bill Hostmann. Magic quadrant for business intelligence platforms. *Gartner Group, Stamford, CT*, 2011.
2. Beyond the pc. Special Report on Personal TEchnology, 2011.
3. Stephen A Goff, Matthew Vaughn, Sheldon McKay, Eric Lyons, Ann E Stapleton, Damian Gessler, Naim Matasci, Liya Wang, Matthew Hanlon, Andrew Lenards, et al. The iplant collaborative: cyberinfrastructure for plant biology. *Frontiers in plant science*, 2, 2011.
4. D Agrawal, P Bernstein, E Bertino, S Davidson, U Dayal, M Franklin, J Gehrke, L Haas, A Halevy, J Han, et al. Challenges and opportunities with big data. a community white paper developed by leading researchers across the united states, 2012.
5. George K Baah, Alexander Gray, and Mary Jean Harrold. On-line anomaly detection of deployed software: a statistical machine learning approach. In *Proceedings of the 3rd international workshop on Software quality assurance*, pages 70–77. ACM, 2006.
6. Michael Moeng and Rami Melhem. Applying statistical machine learning to multicore voltage & frequency scaling. In *Proceedings of the 7th ACM international conference on Computing frontiers*, pages 277–286. ACM, 2010.
7. Mohamed Medhat Gaber, Arkady Zaslavsky, and Shonali Krishnaswamy. Mining data streams: a review. *ACM Sigmod Record*, 34(2):18–26, 2005.

8. Vassilios S Verykios, Elisa Bertino, Igor Nai Fovino, Loredana Parasiliti Provenza, Yucel Saygin, and Yannis Theodoridis. State-of-the-art in privacy preserving data mining. *ACM Sigmod Record*, 33(1):50–57, 2004.
9. Wil van der Aalst. Process mining: Overview and opportunities. *ACM Transactions on Management Information Systems (TMIS)*, 3(2):7, 2012.
10. Chris Ding, Xiaofeng He, Parry Husbands, Hongyuan Zha, and Horst D Simon. Pagerank, hits and a unified framework for link analysis. In *Proceedings of the 25th annual international ACM SIGIR conference on Research and development in information retrieval*, pages 353–354. ACM, 2002.
11. Christopher D Manning and Hinrich Schütze. *Foundations of statistical natural language processing*, volume 999. MIT Press, 1999.
12. Alan Ritter, Sam Clark, Oren Etzioni, et al. Named entity recognition in tweets: an experimental study. In *Proceedings of the Conference on Empirical Methods in Natural Language Processing*, pages 1524–1534. Association for Computational Linguistics, 2011.
13. Yanpeng Li, Xiaohua Hu, Hongfei Lin, and Zhiahi Yang. A framework for semisupervised feature generation and its applications in biomedical literature mining. *IEEE/ACM Transactions on Computational Biology and Bioinformatics (TCBB)*, 8(2):294–307, 2011.
14. David M Blei. Probabilistic topic models. *Communications of the ACM*, 55(4):77–84, 2012.
15. Helen Balinsky, Alexander Balinsky, and Steven J Simske. Automatic text summarization and small-world networks. In *Proceedings of the 11th ACM symposium on Document engineering*, pages 175–184. ACM, 2011.
16. Meenakshi Mishra, Jun Huan, Said Bleik, and Min Song. Biomedical text categorization with concept graph representations using a controlled vocabulary. In *Proceedings of the 11th International Workshop on Data Mining in Bioinformatics*, pages 26–32. ACM, 2012.
17. Jian Hu, Lujun Fang, Yang Cao, Hua-Jun Zeng, Hua Li, Qiang Yang, and Zheng Chen. Enhancing text clustering by leveraging wikipedia semantics. In *Proceedings of the 31st annual international ACM SIGIR conference on Research and development in information retrieval*, pages 179–186. ACM, 2008.
18. Mark T Maybury. *New directions in question answering*. AAAI press Menlo Park, 2004.
19. Bo Pang and Lillian Lee. Opinion mining and sentiment analysis. *Foundations and trends in information retrieval*, 2(1–2):1–135, 2008.
20. Sankar K Pal, Varun Talwar, and Pabitra Mitra. Web mining in soft computing framework: Relevance, state of the art and future directions. *Neural Networks, IEEE Transactions on*, 13(5):1163–1177, 2002.
21. Soumen Chakrabarti. Data mining for hypertext: A tutorial survey. *ACM SIGKDD Explorations Newsletter*, 1(2):1–11, 2000.
22. Sergey Brin and Lawrence Page. The anatomy of a large-scale hypertextual web search engine. *Computer networks and ISDN systems*, 30(1):107–117, 1998.
23. David Konopnicki and Oded Shmueli. W3qs: A query system for the world-wide web. In *VLDB*, volume 95, pages 54–65, 1995.
24. Soumen Chakrabarti, Martin Van den Berg, and Byron Dom. Focused crawling: a new approach to topic-specific web resource discovery. *Computer Networks*, 31(11):1623–1640, 1999.
25. Duo Ding, Florian Metze, Shourabh Rawat, Peter Franz Schulam, Susanne Burger, Ehsan Younessian, Lei Bao, Michael G Christel, and Alexander Hauptmann. Beyond audio and video retrieval: towards multimedia summarization. In *Proceedings of the 2nd ACM International Conference on Multimedia Retrieval*, page 2. ACM, 2012.
26. Meng Wang, Bingbing Ni, Xian-Sheng Hua, and Tat-Seng Chua. Assistive tagging: A survey of multimedia tagging with human-computer joint exploration. *ACM Computing Surveys (CSUR)*, 44(4):25, 2012.
27. Michael S Lew, Nicu Sebe, Chabane Djeraba, and Ramesh Jain. Content-based multimedia information retrieval: State of the art and challenges. *ACM Transactions on Multimedia Computing, Communications, and Applications (TOMCCAP)*, 2(1):1–19, 2006.

28. Weiming Hu, Nianhua Xie, Li Li, Xianglin Zeng, and Stephen Maybank. A survey on visual content-based video indexing and retrieval. *Systems, Man, and Cybernetics, Part C: Applications and Reviews, IEEE Transactions on*, 41(6):797–819, 2011.
29. You-Jin Park and Kun-Nyeong Chang. Individual and group behavior-based customer profile model for personalized product recommendation. *Expert Systems with Applications*, 36(2):1932–1939, 2009.
30. Ana Belén Barragáns-Martínez, Enrique Costa-Montenegro, Juan C Burguillo, Marta Rey-López, Fernando A Mikic-Fonte, and Ana Peleteiro. A hybrid content-based and item-based collaborative filtering approach to recommend tv programs enhanced with singular value decomposition. *Information Sciences*, 180(22):4290–4311, 2010.
31. Milind Naphade, John R Smith, Jelena Tesic, Shih-Fu Chang, Winston Hsu, Lyndon Kennedy, Alexander Hauptmann, and Jon Curtis. Large-scale concept ontology for multimedia. *Multimedia, IEEE*, 13(3):86–91, 2006.
32. Zhigang Ma, Yi Yang, Yang Cai, Nicu Sebe, and Alexander G Hauptmann. Knowledge adaptation for ad hoc multimedia event detection with few exemplars. In *Proceedings of the 20th ACM international conference on Multimedia*, pages 469–478. ACM, 2012.
33. Jorge E Hirsch. An index to quantify an individual's scientific research output. *Proceedings of the National academy of Sciences of the United States of America*, 102(46):16569, 2005.
34. Duncan J Watts. *Six degrees: The science of a connected age*. WW Norton & Company, 2004.
35. Charu C Aggarwal. *An introduction to social network data analytics*. Springer, 2011.
36. Salvatore Scellato, Anastasios Noulas, and Cecilia Mascolo. Exploiting place features in link prediction on location-based social networks. In *Proceedings of the 17th ACM SIGKDD international conference on Knowledge discovery and data mining*, pages 1046–1054. ACM, 2011.
37. Akira Ninagawa and Koji Eguchi. Link prediction using probabilistic group models of network structure. In *Proceedings of the 2010 ACM Symposium on Applied Computing*, pages 1115–1116. ACM, 2010.
38. Daniel M Dunlavy, Tamara G Kolda, and Evrim Acar. Temporal link prediction using matrix and tensor factorizations. *ACM Transactions on Knowledge Discovery from Data (TKDD)*, 5(2):10, 2011.
39. Jure Leskovec, Kevin J Lang, and Michael Mahoney. Empirical comparison of algorithms for network community detection. In *Proceedings of the 19th international conference on World wide web*, pages 631–640. ACM, 2010.
40. Nan Du, Bin Wu, Xin Pei, Bai Wang, and Liutong Xu. Community detection in large-scale social networks. In *Proceedings of the 9th WebKDD and 1st SNA-KDD 2007 workshop on Web mining and social network analysis*, pages 16–25. ACM, 2007.
41. Sanchit Garg, Trinabh Gupta, Niklas Carlsson, and Anirban Mahanti. Evolution of an online social aggregation network: an empirical study. In *Proceedings of the 9th ACM SIGCOMM conference on Internet measurement conference*, pages 315–321. ACM, 2009.
42. Miltiadis Allamanis, Salvatore Scellato, and Cecilia Mascolo. Evolution of a location-based online social network: analysis and models. In *Proceedings of the 2012 ACM conference on Internet measurement conference*, pages 145–158. ACM, 2012.
43. Neil Zhenqiang Gong, Wenchang Xu, Ling Huang, Prateek Mittal, Emil Stefanov, Vyas Sekar, and Dawn Song. Evolution of social-attribute networks: measurements, modeling, and implications using google+. In *Proceedings of the 2012 ACM conference on Internet measurement conference*, pages 131–144. ACM, 2012.
44. Elena Zheleva, Hossam Sharara, and Lise Getoor. Co-evolution of social and affiliation networks. In *Proceedings of the 15th ACM SIGKDD international conference on Knowledge discovery and data mining*, pages 1007–1016. ACM, 2009.
45. Jie Tang, Jimeng Sun, Chi Wang, and Zi Yang. Social influence analysis in large-scale networks. In *Proceedings of the 15th ACM SIGKDD international conference on Knowledge discovery and data mining*, pages 807–816. ACM, 2009.

46. Yanhua Li, Wei Chen, Yajun Wang, and Zhi-Li Zhang. Influence diffusion dynamics and influence maximization in social networks with friend and foe relationships. In *Proceedings of the sixth ACM international conference on Web search and data mining*, pages 657–666. ACM, 2013.
47. Theodoros Lappas, Kun Liu, and Evimaria Terzi. Finding a team of experts in social networks. In *Proceedings of the 15th ACM SIGKDD international conference on Knowledge discovery and data mining*, pages 467–476. ACM, 2009.
48. Tong Zhang, Alexandrin Popescul, and Byron Dom. Linear prediction models with graph regularization for web-page categorization. In *Proceedings of the 12th ACM SIGKDD international conference on Knowledge discovery and data mining*, pages 821–826. ACM, 2006.
49. Yang Zhou, Hong Cheng, and Jeffrey Xu Yu. Graph clustering based on structural/attribute similarities. *Proceedings of the VLDB Endowment*, 2(1):718–729, 2009.
50. Wenyuan Dai, Yuqiang Chen, Gui-Rong Xue, Qiang Yang, and Yong Yu. Translated learning: Transfer learning across different feature spaces. In *Advances in Neural Information Processing Systems*, pages 353–360, 2008.
51. Cisco Visual Networking Index. Global mobile data traffic forecast update, 2012–2017 http://www.cisco.com/en.US/solutions/collateral/ns341/ns525/ns537/ns705/ns827/white_paper_c11-520862.html(Sonerişim:5May\T1\is2013), 2013.
52. Youngho Rhee and Juyeon Lee. On modeling a model of mobile community: designing user interfaces to support group interaction. *interactions*, 16(6):46–51, 2009.
53. Jiawei Han, Jae-Gil Lee, Hector Gonzalez, and Xiaolei Li. Mining massive rfid, trajectory, and traffic data sets. In *Proceedings of the 14th ACM SIGKDD international conference on Knowledge discovery and data mining*, page 2. ACM, 2008.
54. Eugene Wu, Yanlei Diao, and Shariq Rizvi. High-performance complex event processing over streams. In *Proceedings of the 2006 ACM SIGMOD international conference on Management of data*, pages 407–418. ACM, 2006.
55. Manoj K Garg, Duk-Jin Kim, Deepak S Turaga, and Balakrishnan Prabhakaran. Multimodal analysis of body sensor network data streams for real-time healthcare. In *Proceedings of the International Conference on Multimedia information retrieval*, pages 469–478. ACM, 2010.
56. Yubin Park and Joydeep Ghosh. A probabilistic imputation framework for predictive analysis using variably aggregated, multi-source healthcare data. In *Proceedings of the 2nd ACM SIGHIT International Health Informatics Symposium*, pages 445–454. ACM, 2012.
57. Viktor Mayer-Schönberger and Kenneth Cukier. *Big Data: A Revolution that Will Transform how We Live, Work, and Think.* Eamon Dolan/Houghton Mifflin Harcourt, 2013.

Chapter 7
Open Issues and Outlook

Abstract In the previous chapters, we review the background and state-of-the-art of big data. In Fig. 7.1, it illustrates all the key technologies of big data introduced in this book. In this chapter, we summarize the research hot spots and suggest possible research directions of big data. We also discuss potential development trends in this broad research and application area.

7.1 Open Issues

The analysis of big data is confronted with many challenges but the current research is still in the beginning phase. Considerable research efforts are needed to improve the efficiency of data display, data storage, and data analysis.

7.1.1 Theoretical Research

Although big data is a hot research area in both academia and industry, there are many important problems remain to be solved, which are discussed below.

- *Fundamental Problems*: There is compelling need for a rigorous definition of big data, a structural model of big data, formal description of big data, and a theoretical system of data science, etc. At present, many discussions of big data look more like commercial speculation than scientific research. This is because big data is not formally and structurally defined and not strictly verified.
- *Standardization*: An evaluation system of data quality and an evaluation standard of data computing efficiency should be developed. Many solutions of big data applications claim they can improve data processing and analysis capacities in all aspects, but there is still not a unified evaluation standard and benchmark to balance the computing efficiency of big data with rigorous mathematical

M. Chen et al., *Big Data: Related Technologies, Challenges and Future Prospects*,
SpringerBriefs in Computer Science, DOI 10.1007/978-3-319-06245-7_7,

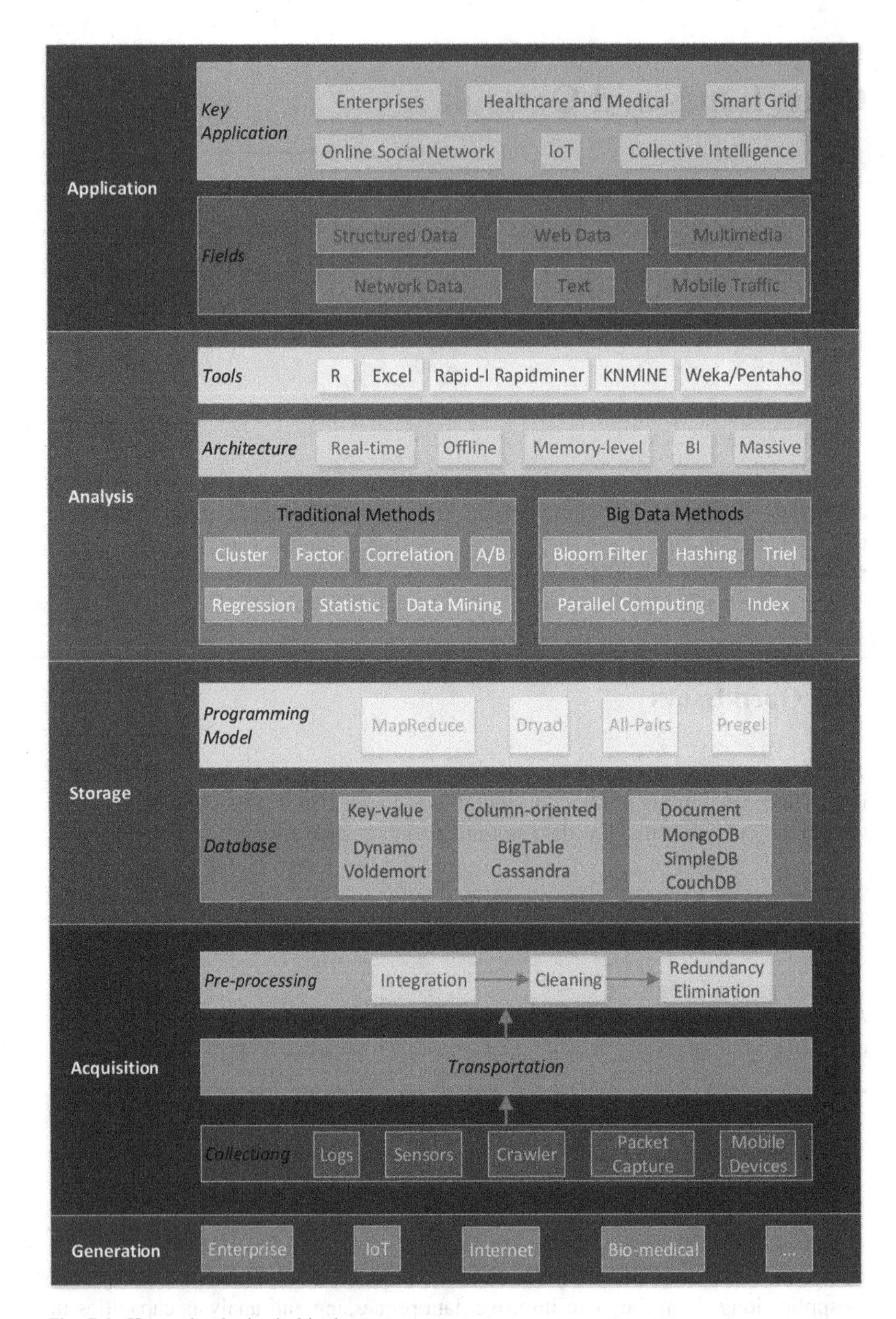

Fig. 7.1 Key technologies in big data era

methods. The performance can only be evaluated by the system is implemented and deployed, which could not horizontally compare advantages and disadvantages of various solutions and compare efficiencies before and after the use of big data. In addition, since data quality is an important basis of data preprocessing, simplification, and screening, it is also an urgent problem to effectively evaluate data quality.

- *Evolution of Big Data Computing Modes*: This includes external storage mode, data flow mode, PRAM mode, and MR mode, etc. The emergence of big data triggers the development of algorithm design, which has transformed from a computing-intensive approach into a data-intensive approach. Data transfer has been a main bottleneck of big data computing. Therefore, many new computing models tailored for big data have emerged and more such models are on the horizon.

7.1.2 Technology Development

The big data technology is still in its infancy. Many key technical problems, such as cloud computing, grid computing, stream computing, parallel computing, big data architecture, big data model, and software systems supporting big data, etc. should be fully investigated.

- *Format Conversion*: Due to wide and various data sources, heterogeneity is always a characteristic of big data, as well as a key factor which restricts the efficiency of data format conversion. If such format conversion can be made more efficient, the application of big data may create more values.
- *Big Data Transfer*: Big data transfer involves big data generation, acquisition, transmission, storage, and other data transformations in the spatial domain. As discussed, big data transfer usually incurs high costs, which is also the bottleneck for big data computing. However, data transfer is inevitable in big data applications. Improving the transfer efficiency of big data is a key factor to improve big data computing.
- *Real-time Performance*: The real-time performance of big data is also a core problem in many different application scenarios. Ways to define the life cycle of data, compute the rate of depreciation of data, and build computing models of real-time applications and online applications, will influence the values and analytical and feedback results of big data.

As big data research is advanced, new problems on big data processing arise from the traditional simple data analysis, including: (a) data re-utilization, since big data features big value but low density, with the increase of data scale, more values may be mined from re-utilization of existing data; (b) data re-organization, datasets in different businesses can be re-organized, with the total re-organized data values larger than the total datasets' value; (c) data exhaust, unstructured information or data that is a by-product of the online activities of Internet users. In big data, not only correct data should be utilized, but also the wrong data should be utilized to generate more value. Collecting and analyzing data exhaust can provide valuable insight into the purchasing habits of consumers.

7.1.3 Practical Implications

Although there are already many successful big data applications, many practical problems should be solved:

- *Big Data Management*: the emergence of big data brings about new challenges to traditional data management. At present, many research efforts are being made on consider big data oriented database and Internet technologies, management of storage models and databases of new hardware, heterogeneous and multi-structured data integration, data management of mobile and pervasive computing, data management of SNS, and distributed data management.
- *Searching, Mining, and Analysis of Big Data*: data processing is always a research hotspot in the big data field, e.g., searching and mining of SNS models, big data searching algorithms, distributed searching, P2P searching, visualized analysis of big data, massive recommendation systems, social media systems, real-time big data mining, image mining, text mining, semantic mining, multi-structured data mining, and machine learning, etc.
- *Integration and Provenance of Big Data*: As discussed, the value acquired from a comprehensive utilization of multiple datasets is higher than the total value of individual datasets. Therefore, the integration of different data sources is a timely problem to be solved. Data integration is to integrate different datasets from different sources, which are confronted with many challenges, such as different data patterns and large amount of redundant data. Data provenance is to describe the process of data generation and evolution over time. In the big data era, data provenance is mainly used to investigate multiple datasets other than a single dataset. Therefore, it is worth of study on how to integrate data provenance information featuring different standards and from different datasets.
- *Big Data Application*: at present, the application of big data is just beginning and we shall explore and more efficiently ways to fully utilize big data. Therefore, big data applications in science, engineering, medicine, medical care, finance, business, law enforcement, education, transportation, retail, and telecommunication, big data applications in small and medium-sized businesses, big data applications in government departments, big data services, and human-computer interaction of big data, etc. are all important research problems.

7.1.4 Data Security

In IT, safety and privacy are always two key concerns. In the big data era, as data volume is fast growing, there are more severe safety risks, while the traditional data

protection methods have already been shown not applicable to big data. In particular, big data safety is confronted with the following security related challenges.

- *Big Data Privacy*: In the big data era, data privacy includes two aspects: (a) the protection of personal privacy, as the advances on data acquisition is made, personal interests, habits, and body properties, etc. of users may be more easily acquired, of which the user may not be aware. (b) Personal privacy data may also be leaked during storage, transmission, and usage, even if acquired with the permission of users. Facebook is deemed as a big data company with the most SNS data currently. Organizations that own big data usually attempt to mine valuable information in the data with advanced algorithms. The privacy data protection technology therefore is of great importance. According to a report [1], Ron Bowes, a researcher of Skull Security, acquired data in the public pages of Facebook users who fail to modify their privacy setting using an information acquisition tool. Ron Bowes packaged such data into a 2.8 GB package and created a BT seed for others to download. The analysis capacity of big data may lead to privacy mining from seemingly simple information. Therefore, privacy protection in the big data era will become a new and challenging problem.
- *Data Quality*: Data quality influences big data utilization. Low quality data wastes transmission and storage resources, and may not be usable. There are a lot of factors that may restrict data quality, for example, generation, acquisition, transmission, and transmission may all influence data quality. Data quality is mainly manifested in its accuracy, completeness, redundancy, and consistency. Even though a lot of measures have been taken to improve data quality, the quality related problems could not be completely solved. Therefore, effective methods to automatically detect data quality and repair some damaged data need to be investigated.
- *Big Data Safety Mechanism*: Big data brings challenges to data encryption due to its large scale and high variety. The performance of previous encryption methods on small and medium-scale data could not meet the demands of big data; efficient big data cryptography approaches shall be developed. Effective schemes for safety management, access control, and safety communications shall be investigated for structured, semi-structured, and unstructured data. In addition, under the multi-tenant mode, isolation, confidentiality, completeness, availability, controllability, and traceability of tenants' data should be enabled on the premise of efficiency assurance.
- *Big Data Application in Information Security*: Big data not only brings challenges to information security, but also offers new opportunities for the development of information security mechanisms. For example, we may discover potential safety loopholes and APT (Advanced Persistent Threat) after the analysis of the big data in the form of log files of an Intrusion Detection System. In addition, virus characteristics, loophole characteristics, and attack characteristics, etc. may also be more easily identified through the analysis of big data.

To sum up, the safety of big data has drawn great attention of researchers. However, there is only limited research on the representation of multi-source heterogeneous big data, measurement and semantic comprehension methods, modeling theories and computing models, distributed storage of energy efficiency optimization, and processed hardware and software system architectures, etc. Particularly, big data security, including big data credibility, big data backup and recovery technologies in various application fields, big data completeness maintenance technology, and big data security technology should be further investigated.

7.2 Outlook

The emergence of big data opens great opportunities. In the IT era, the "T" (Technology) was the main concern, while technology derives the development of data. In the big data era, with the prominence of data value and the advances in I (Information), data will drive the progress of technologies in the future. Big data will not only change the social and economic life, but also influence everyone's ways of living and thinking, which is just beginning. We could not predict the future but may take precautions for possible events to occur in the future.

- *Data With a Larger Scale, More Variety, and More Complex Structures*: Although technologies represented by Hadoop have achieved a great success, such technologies are definitely to fall behind and will be replaced given the rapid development of big data. For example, the theoretical basis of Hadoop has emerged as early as 2006. Many researchers have concerned ways to better cope with larger-scale, more various kinds of, and more complexly structured data. These efforts are represented by (Globally-Distributed Database) Spanner of Google and fault-tolerant and expandable distributed relational database F1. In the future, the storage technology of big data will be based on distributed databases, support transaction mechanisms similar to relational databases, and effectively handle data through grammars similar to SQL.
- *Data Resource Performance*: Since big data contains huge values, mastering big data means mastering resources. Through the analysis of the value chain of big data, it can be seen that its value comes from the data itself, technologies, and ideas, and the core is data resources. Without data technologies and ideas, values could not be created. The reorganization and integration of different datasets can create more values. From now on, enterprises that master big data resources may obtain huge benefits by renting and assigning the rights to use their data.
- *Big Data Promotes the Cross Fusion of Science*: Big data not only promotes the comprehensive fusion of cloud computing, Internet of Things, data center, and mobile networks, etc., but also forces the cross fusion of many disciplines. The development of big data shall explore innovative technologies and methods in big data acquisition, storage, processing, mining, and information security, etc., based on information science, and examine changes and impacts of big data on

production management, business operation and decision making, etc. of modern enterprises from the management perspective. What's more, the application of big data to specific fields needs the participation of interdisciplinary talents.

- *Visualization*: In many human-computer interaction scenarios, the principle of What You See Is What You Get is followed, e.g., text and image editors. In big data applications, mixed data may not be is very useful for decision making. Only when the analytical results are friendly displayed, it may be accepted and utilized by users. Reports, histograms, pie charts, and regression curves, etc., are frequently used to visualize results of data analysis. New presentation forms will occur in the future, e.g., Microsoft Renlifang, a social search engine, utilizes relational diagrams to express interpersonal relationship.
- *Data-Oriented*: It is well-known that programs are consisted of data structures and algorithms. In the history of program design, it is observed that the role of data is becoming increasingly more significant. In the small scale data era, in which logic is more complex than data, program design is mainly focused on processes. As business data is becoming more complex, object-oriented design methods are developed. The complexity of business data has far surpassed business logic and programs gradually transform from algorithm-intensive to data-intensive. It is anticipated data-oriented program design methods are certain to emerge, which will have far-reaching influence on the development of IT in software engineering, architecture, and model design, among others.
- *Big Data Causes the Revolution of Thinking*: In the big data era, data collection, acquisition, and analysis are more rapidly accomplished and the massive data will profoundly influence our ways of thinking. In [2], the authors summarizes the thinking revolution caused by big data as follows:

  - During data analysis, we will try to utilize all data other than only analyzing a little sample data.
  - Compared with accurate data, we would like to accept numerous and complicated data.
  - We shall pay greater attention to correlations between things other than exploring causal relationship.
  - The simple algorithms of big data are more effective than complex algorithms of small data.
  - Analytical results of big data will reduce hasty and subjective factors during decision making and data scientists will replace "experts."

- *Managing Large-scale FlowTable for Software-Defined Networking with Big Data Techniques*: In the past few years, software-defined networking (SDN) has been the buzz of the networking world. It was originally proposed to accelerate networking innovations in legacy campus networks called OpenFlow, which comprises a number of closed networking boxes with diverse functionalities (such as routing, switching, firewall, etc.) [3]. It is observed that, plenty of emerging networking problems appeared in the era when cloud computing meets big data applications, and SDN seems to be extremely suitable for solving those problems in respect of network efficiency, scalability, flexibility, agility, as well as operation and maintenance complexity. In the specification of OpenFlow, one of the most important concept is FlowTable, which includes a large number of rules to process network packets. Obviously, it is a challenge to manage the large-scale FlowTables. A promising way is to implement SDN with big data techniques, to effectively store, process and utilize FlowTable, and increase the speed of searching rules.

- *5G Wireless Networks: Supporting Technology for Mobile Big Data*: With the emergence of cloud computing as an important information technology in support of virtualized services, it becomes promising to design 5G wireless networks by exploiting recent advances relevant to network function virtualization and benefiting from advanced virtualization techniques of cloud computing to build efficient and scalable networking infrastructures. Researchers have been designing new architectures for elastically composing and operating a virtual end-to-end network platform on demand on top of fragmented physical infrastructures provided by federated cloud. SDN techniques have been seen as promising enablers for this vision of carrier cloud, which will likely play a crucial role in the design of 5G wireless networks.

  Due to the huge explosion in mobile data of a hyperconnected society, "Can Big Data go Mobile?" now becomes a challenging problem which would be addressed by 5G technologies. Though 5G wireless provides the possibility to enable the mobility of big data, there are various research problems towards the realization of the brand-new networking system, such as 5G network architecture, SDN and network virtualization techniques for enabling 5G, resource allocation algorithms in 5G, and 5G-related control protocols and optimization techniques. In an energy efficient, flexible, connectivity-scalable and secure manner, new business models beyond IaaS, PaaS and SaaS, such as Network as a Service (NaaS), and Knowledge as a Service (KaaS), are expected to emerge. Especially, Big Data as a Service (BDaaS) or Big Data Analysis as a Service (BDAaaS) could emerge, facilitating the efficient storage and analysis for the exploding mobile data.

Throughout the history of human society, the demands and willingness of human beings are always the source powers to promote scientific and technological progress. In the big data era, big data may provides reference answers for human beings to make decisions through mining and analytical processing, but could not replace human thinking. It is human thinking that promotes the widespread utilizations of big data. Big data is more like an extendable and expandable human brain other than a substitute of human brain. With the emergence of Internet of Things, development of mobile sensing technology, and progress of data acquisition technology, people are not only the user and consumer of big data, but also its producer and participant. Social relation sensing, crowdsourcing, analysis of big data in SNS, and other applications closely related to human activities based on big data will be increasingly concerned and will certainly cause enormous changes of social activities in the future society.

References

1. Predrag Tasevski. Password attacks and generation strategies. *Tartu University: Faculty of Mathematics and Computer Sciences*, 2011.
2. Viktor Mayer-Schönberger and Kenneth Cukier. *Big Data: A Revolution that Will Transform how We Live, Work, and Think*. Eamon Dolan/Houghton Mifflin Harcourt, 2013.
3. McKeown, Nick and Anderson, Tom and Balakrishnan, Hari and Parulkar, Guru and Peterson, Larry and Rexford, Jennifer and Shenker, Scott and Turner, Jonathan. OpenFlow: enabling innovation in campus networks. *ACM SIGCOMM Computer Communication Review*, 38(2):69–77, 2008.